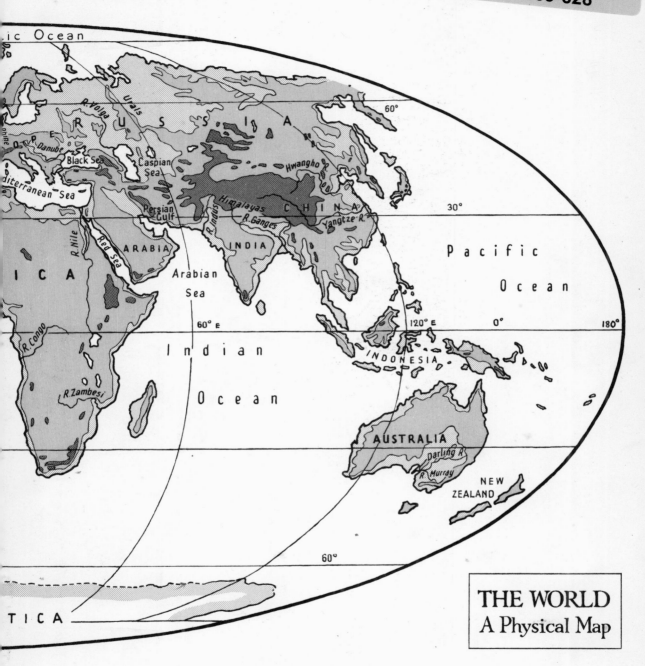

ic Ocean

R.Volga

Urals

R U S S I A

60°

E R

R.Danube

Black Sea

Caspian
Sea

Hwangho

diterranean Sea

Persian
Gulf

Himalayas

C H I N A

30°

R.Nile

Red Sea

ARABIA

R.Indus

R.Ganges

Yangtze R.

I C A

INDIA

Pacific

Arabian

Ocean

Sea

60° E

0°

180°

I n d i a n

INDONESIA

120° E

R.Congo

O c e a n

R.Zambesi

AUSTRALIA

Darling R.

R. Murray

NEW
ZEALAND

60°

TICA

THE WORLD
A Physical Map

LOOKING AT THE WORLD

A street on the Greek island of Seriphos

General Editor: R. J. Unstead

Looking at
THE WORLD

JEAN and DAVID GADSBY

Adam and Charles Black · London

Originally published in four parts with the title
LOOKING AT GEOGRAPHY
Part 4 was written in association with G. M. Ashby

ISBN 0 7136 1276 2

FOURTH EDITION 1971

First edition 1961 Second edition 1964 Third edition 1967
© 1971 A & C BLACK LTD 4, 5 & 6 SOHO SQUARE
LONDON W1V 6AD
MADE AND PRINTED IN GREAT BRITAIN BY
J W ARROWSMITH LTD, BRISTOL

INTRODUCTION

This is a book about the world today. It shows how people live in every kind of land, from the hot deserts of Brazil to the frozen wastes of Russia, from the deserts of Australia to the crowded cities of North America.

It has been written for children and it contains more than 1000 pictures. These include photographs, drawings, maps and diagrams, each one carefully chosen or specially drawn for this book.

The book is in four parts:

Part 1 tells of the lives of children of other lands

Part 2 tells of things we eat and use in our daily lives

Part 3 tells of Britain today

Part 4 shows how people live in parts of the world which differ
        widely in climate and development.

Throughout the four parts there are chapters about maps, the weather, the seasons, and travel in the world today.

Israel

CONTENTS

A complete list of the contents of each part will be found at
the beginning of each of the four parts
A summary of the contents of the complete book is given on pages viii to xi

Ceylon

viii   *Jerusalem*

Venice

Australia

Miami, Florida

Andes, South America

x

A weather ship

Wales

PHYSICAL GEOGRAPHY

LET US REMEMBER

The earth

COLOUR ILLUSTRATIONS

Acknowledgements

Acknowledgement is gratefully made to the following for the use of colour photographs:
1 N. Kontos; 2 and 3 Normanns Kunstforlag A/S; 4 L. van Leer & Co NV and KLM Aerocarto NV; 5 L. van Leer & Co NV; 6 and 7 Jonathan Rutland; 8 The Observer; 9 Controller of Her Majesty's Stationery Office (Crown Copyright); 10 Jonathan Rutland; 11 and 12 Gwynneth Ashby.

Acknowledgement is also due to the following for the use of photographs:
Australian News and Information Bureau ix b; Barnaby's Photo Library xi b; British Antarctic Survey viii a; Camera Press v; J. Allan Cash viii c; Ceylon Tea Centre viii b; Fotofilm ix a; Meteorological Office xi a; Popperfoto x c; Jonathan Rutland vii; US Information Service x a and b, xi c

PART ONE

Looking at
Other Children

The publishers are grateful to the Association of Agriculture
for allowing them to base the farm picture and plan on
page 35 on one of the farms in their Adoption Scheme.
They are also grateful to the following for permission to
reproduce photographs:
Automobile Association   7a
Barnaby's Picture Library   1, 9, 12, 13b, 15, 17, 19a, 23, 25,
27, 30, 37b and c, 41a and b, 53a and b, 55, 58a and b
Bergen Tourist Office   49, 52b
Camera Press   3, 13a, 20, 43, 48a and b
Farmer and Stockbreeder   37a, 40
Fox Photos   6, 11a and b
International Harvester Co of Great Britain   38
KLM Aerocarto NV   60
Landbrukets Sentralforbund   51
Massey-Harris-Ferguson Ltd   5a, 39c
Norway Export Council   52a
Popperfoto   5b, 7b, 52c, 57
Rediffusion   34
Royal Danish Embassy   19b, 24b
Royal Danish Ministry for Foreign Affairs.   24a
Royal Netherlands Embassy   61, 62a and b
Sport and General   33
The Times   42a and b

CONTENTS OF PART 1

About part 1

This part is about the children of many lands. It tells you what they wear and what they eat; it tells you about their homes and about the work their fathers do.

One of these children lives in a hot, wet forest; another lives in a very cold land. One child's home is a tent; another lives in a house made of logs.

One child lives in a house made of bamboo and mud; an ox pulls his father's plough. Another child lives in a brick house, and his father uses a tractor to pull his plough. The father of one of these children grows rice; another grows flower bulbs.

In this part you will also read about the seasons, and shadows, about the sun, and the rain, and you will learn how to draw plans.

The Seasons

Autumn

In the early autumn the farmers are busy with the harvest. When the ripe corn is being cut everyone hopes for fine weather.

Combine harvesters cut the corn and thresh it.

The grain drops from the arm on top of the machine into the container at the back. The chaff and straw are blown back on to the field.

From corn we get flour to make bread and biscuits. Porridge, and food for the farm animals are made from corn.

The leaves on the trees turn red, yellow and brown in the autumn and fall to the ground. In the town the roadmen work hard to sweep them up, so that they do not block the gutters.

Some of the trees keep their leaves all the year round. They are called evergreens. Holly and fir trees are evergreen.

5

Autumn

In the autumn many kinds of fruit are ripe and ready for eating. Apples and pears are picked in the orchards. They are packed in boxes and sent to market.

"Conkers" are ripe on the horse-chestnut trees, and children gather ripe blackberries in the woods and hedges.

We see less and less of the sun each day, and so the days grow cooler. At home we need the lights on earlier every day.

In the parks the keepers dig up the dead flowers and burn them. Some plants are split up so that they will flower well next year.

Sometimes there are thick fogs, and cars need their lights on even during the day.

Blackberries

Winter

·In winter it is dark when we get up in the morning, and dark when we go to bed. The fields look very bare. The grass is too short for the cows to eat, and the farmers feed them with hay. Car and lorry drivers have to be careful when the roads are icy. Sometimes sand is thrown on the roads to make them safer.

Robins and sparrows are very tame, and they wait near our houses for crumbs. You can help to feed the birds by putting out bread and fresh water every day.

In the country the snow often piles up in drifts, and the roads are blocked.

Sometimes the snowploughs cannot clear the roads, and aeroplanes or helicopters have to take food to the villages which are cut off.

Spring

Spring is always an exciting time. The days grow longer, for the sun climbs higher in the sky every day. Soon you can play outdoors after tea. Father digs the garden. He plants onion, lettuce, and carrot seed, and sows early peas and flower seeds.

Birds build their nests and lay their eggs. When the chicks hatch out the birds fly busily to and fro carrying food to them.

The crocuses and snowdrops are the first flowers to open. Soon the primroses and daffodils are in flower, and the leaves of the horse-chestnut trees begin to unfold.

Mother notices how dusty everywhere looks in the sunlight. She sets to work "spring-cleaning"—scrubbing, dusting and polishing every room.

Summer

In summer the days are long. The farmer cuts his hay, and we all want to be in the country, or paddling and swimming at the seaside.

The bees are busy in the gardens, gathering nectar to make honey.

In the towns the shopkeepers pull down their blinds, so that the sun does not fade the brightly coloured hats and dresses in their windows.

Seasons and Shadows

This drawing shows you when the seasons are.

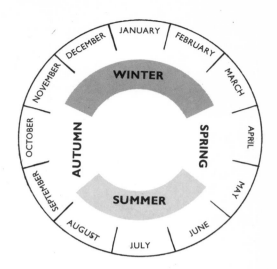

Which is the longest season?

Which are the best months for going to the seaside?

Which are the autumn months?

If you want to know where north and south are, at any time of the year, there is an easy way of finding out in the sunshine. Go out at midday. If you stand with your back to the sun your shadow is pointing north.

If you stretch your arms out sideways, your left arm points to the west, your right arm to the east. The south is behind you.

But there is a difference between summer and winter shadows. Can you see what it is?

Winter

Summer

10

Here is a farm in the middle of winter.

The shadows are long because the sun is low in the sky.

The day is bright and clear, but the sun is too low in the sky to melt the snow.

There are no leaves on the big elm trees.

The second picture shows the same farm in summer. Now the sun is high in the sky, and the shadows are short.

The warm sun has made the nettles and grass grow tall.

Both these pictures were taken at noon, so we can tell by the shadows that the north is to the right. The camera is looking west.

11

Rain

You know that rain comes from the clouds. But do you know that the clouds are really made up of millions of very tiny drops of water? If the drops grow bigger they become too heavy to stay in the air, and they fall as rain. If it is very cold, the raindrops turn to snow.

We often think rain is a nuisance, but we could not live without it. We need water to drink, and so do the cows which give us milk and meat.

The fruit and vegetables which we eat must have water to make them grow.

Many countries have far more rain than Britain. They have great forests and wide rivers. Some countries have hardly any rain, and no trees or grass can grow there. The people who live in these countries have a very hard life, looking for water, and grass for their animals.

12

Peko of the Amazon Forest

The River Amazon is in South America. Hundreds of little rivers flow into it, until it becomes the largest river in the world. Peko lives near one of these little rivers.

The forest around Peko's home is so thick that he can hardly see the sun through the trees, and it is so huge that no one has yet explored it all. Creepers as thick as a man's arm twine round the tree trunks, and great trees hide the sun.

Peko wears hardly any clothes as it is hot and steamy in the forest all the year round. There is no winter and no summer, and the trees are always green. In the evening the sun goes down quickly, and suddenly everywhere is dark.

If Peko runs into a tree he may bring down a shower of red ants which will bite him. When Peko is thirsty he cuts a piece of water vine. He holds it so that the sap runs into his mouth.

13

When Peko's father wants to build a new hut he chooses a place not far from the river. But he makes sure that his hut cannot be seen by men coming up the river in boats.

To clear a space, the men of the tribe cut down the trees and creepers with their axes. They leave one tall tree standing, and use it as a centre pole for their new hut. They make a frame of branches round the centre pole, and tie the frame with creepers which are as strong as rope.

They cover the walls and roof of the hut with palm leaves. The doorway is a small opening in the side of the hut. A curtain of leaves hangs over the door.

The hut is dark inside, for it has no windows. The only light comes from the doorway, and from the fires which are always burning.

About sixty people live in Peko's hut. They sleep on the floor, or in hammocks made of twisted palm leaves.

14

Each family has its own fire. There is no chimney, but some of the smoke finds its way out by a small hole in the roof. The smoke helps to keep out the flies and mosquitoes.

The Chief lives on a platform at one side of the hut. In the middle of the hut is a space where the children play.

After two or three years the roof of Peko's hut begins to leak. Then the hut is burned down and Peko and his parents move to another part of the forest, where they build a new hut.

15

When the men have built Peko's new hut, the ground around it is cleared. Then Peko's mother and the other women can grow sugar cane, maize, and *cassava,* which is a kind of vegetable root. Peko's mother collects the large roots of the cassava in a basket on her back.

The roots of the cassava are poisonous, but she peels and grates them and soaks them in water to get rid of the poison. Then she drains them and bakes flat cakes over the fire.

Peko's mother collects nuts from the trees in the forest. She cuts the bark of the rubber trees, and collects the juice which runs from them.

She goes to the river and with the clay which she finds there she makes pots. She also makes baskets from palm leaves.

Peko's mother does not wear many clothes, but she likes to make herself beautiful. She paints patterns on her face and body with a red dye. She wears a necklace made from the teeth of a jaguar.

Peko's father and the men of the tribe are hunters. They shoot poisoned arrows from long blow-pipes and kill the animals and birds of the forest. There are no very large animals in the forest, for it is too thick for them to move about freely.

At night black tapirs come out to find food. The cries of the jaguar echo through the forest.

A parrot and a toucan

Peko is too young to go hunting. But even from the door of his hut he can see monkeys swinging by their tails from the trees.

Peko sees brightly coloured parrots and toucans flying above him. In the bushes there are large hairy spiders. They are so large that they can catch small birds in their webs and kill them.

A monkey

Peko has to be careful of the poisonous snakes on the ground, or he may be bitten.

A tapir

There are no roads in the forest, but there are many rivers because there is so much rain.

The forest is so thick that the only way to travel far is by a river.

An alligator

Peko's father has a large canoe which he hollowed out of a tree trunk. It is very strong, and it takes three men to carry it.

Sometimes Peko goes fishing with his father. They spear turtles and fish, and take them home to be cooked. Peko must be careful, for some of the fish bite, and there are needle fish with sharp points on their backs.

Peko sees alligators basking in the sun on the banks, or sliding into the water.

Few people ever come into the forest, but sometimes traders come up the river. The men and women of Peko's tribe give nuts and rubber to the traders, and the traders give them knives and cloth.

Kara of Greenland

Greenland is a large island which is covered with snow and ice nearly all the year round. Even in summer it is too cold for grass to grow, and there are no trees, only low bushes.

Winter lasts for most of the year, and for three months it is almost dark at midday!

Kara is a little girl who lives in the North-West of Greenland. She has a broad face and slanting eyes, and her hair is straight and black.

It is very cold, so Kara wears plenty of thick clothes. She has two pairs of sealskin boots, fur trousers and a fur hood and mitts. Her fur coat has no buttons. It slips on over her head. Kara's mother made all her clothes. She put the fur side inside so that Kara is always warm. Kara's baby brother keeps warm in his mother's *amaut*, on her back.

19

Kara of Greenland

Kara lives in a small wooden house which has two rooms and a large porch outside. Under the porch supplies of food and tools are kept.

When Kara's mother was a little girl, she lived in a house made from blocks of snow, called an igloo. In summer, when the sun melted the walls of the igloo, she lived in a tent.

Nowadays most families live in houses like Kara's. They have a cooking range, couches to sleep on and chests to sit on and store things in.

This is the inside of Kara's house. Kara's doll is made of sealskin. The puppy will soon be too fierce to live in the house.

Kara's father is putting a new tip on his harpoon.

Sometimes the men go hunting. They shoot polar bears and white foxes.

A polar bear and a white fox

They sell the furs at the trading station, where they buy food, clothes, radios and ammunition for their guns.

Sometimes in the winter there is not enough food stored in the house for Kara and her family. So her father sets off to hunt for seals. He goes on his sledge, pulled by a team of dogs. If he is away from home for more than a day, Kara's father puts up a tent to sleep in.

The sledge is strongly made of wood, and the runners are covered in frozen mud, so that they will slide easily.

The dogs are strong and fierce, but Kara's father controls them with shouts and with his whip.

On the ice Kara's father looks for the breathing-holes of the seals. He builds a wind-shelter from blocks of snow, and settles down to wait. He waits for a long time until the seal comes up to breathe.

Then Kara's father drives his harpoon into it.

The harpoon is tied to his wrist with a long seal-skin rope, and he often has a hard battle before he can pull the seal on to the ice.

A harpoon

The dogs drag the seal home. Kara's father skins the seal. Her mother uses the skin to make tents, ropes and coats. She saves some of the skin for making boats. Sometimes she sells a seal skin at the trading station.

Under the seal's skin is a layer of fat, called blubber. Kara's family eat a great deal of raw blubber to keep themselves warm. They like to drink the warm blood of a seal, because in their country they have no milk.

Kara's father sometimes catches fish. Her mother salts them and hangs them to dry on a sealskin string.

At the end of the winter the snow begins to melt. Soon the rocky ground shows through, and for a short time the ground is covered with a carpet of tiny flowers.

When the ice along the shore begins to break up, Kara's father goes hunting for seals and walrus.

The men set off in their light boats, called *kayaks*. The kayaks have wooden frames and the sides and top are covered with sealskin. When Kara's father is sitting in his kayak he straps himself in tightly, so that no water can get inside it.

At night he sets up a tent made from sealskins or canvas stretched over wooden poles.

A walrus

A kayak is very difficult to paddle, but Kara's father is so clever that he can turn his kayak over and come up again, still strapped in.

Kara of Greenland

When the men return, the seals are skinned. Kara's mother scrapes the flesh from the skin and cuts up the meat and blubber. Some of the skins are taken to the trading station.

Every year there are fewer seals for Kara's father to hunt. Most of his friends now go fishing for cod and shrimps, and they use motor boats instead of kayaks. The fish are put into cans and sent to other countries.

During the day Kara goes to school. She is learning to read and write.

A shrimp packing factory.

Perhaps when Kara is grown up she will work in an office or at the canning factory in the town. Here she is with her friends in the school playground.

Nasir of Saudi Arabia

Nasir lives in Saudi Arabia, which is a hot, dry country where it hardly ever rains. Most of Saudi Arabia is desert, with sand, rocks and prickly bushes.

The people of Nasir's tribe are Bedouin Arabs.

They do not stay in one place, but move about the desert in search of grass and water for their sheep, goats and camels.

Sometimes the Bedouins find a place where there is water and where a few trees grow. It is called an *oasis*.

At the oasis Nasir's father draws water from a well. The animals drink some of the water from a trough. Nasir's father stores the rest in goatskin bags. His buckets are made of camel skin.

Water is very scarce, so an Arab never wastes any. Sometimes he even washes his face in sand, to save his water for drinking.

25

Because Saudi Arabia is a hot country, Nasir wears clothes which help to keep him cool.

He has a white smock with wide sleeves, and a long blue cotton cloak over it. On his head is a dark blue cloth which keeps off the sun. It is held in place by a rope made of camel hair.

Nasir rides on a camel. The camels are very well suited to the desert. They have large feet and they can walk easily on the soft sand.

Their humps are stores of fat. They can live on this fat for many days, even if they have no water to drink.

The skin of a dead camel is made into bags and buckets, and its meat is good to eat.

Nasir's father guards his camels carefully from raiding Arabs. He brands their necks so that he knows which are his. At night he ties their front legs, so that they cannot move.

The Chief of Nasir's tribe rides an Arab horse, which is very fast and strong. The Bedouins are very proud of their famous horses, and Nasir hopes to have his own horse one day.

As the Bedouins journey across the desert, all their tents and baggage are carried on the camels' backs. The shepherds go ahead to look for grass and water. When they have found grass for the animals to eat, one of them rides to tell the Chief. He leads the rest of the tribe, and tells them where to pitch the tents.

The baggage is unloaded from the camels, and Nasir's mother and the other women unroll the tents. They put tent poles under the roof, and fasten the side curtains to keep out the wind.

Nasir helps to drive the tent pegs into the ground, and to pull on the ropes which hold up the tent. He lays the carpets inside the tent, and then their home is ready.

Nasir's tent is made of black goat hair. It is pitched with its back towards the wind. But when the wind changes, the women close one side of the tent and open the other, so that the sand does not blow in.

27

On one side of the tent is the kitchen, and the place where the women live and sleep. Nasir's baby sister sleeps here in a cradle.

There are cooking pots on the floor, and bags of rice and salt. Just outside the women's side of the tent are the water buckets, and a store of wood for the fire.

Nasir lives on the other side of the curtain with his father and the men of the tribe. Here are the saddles and rifles, the fire and the copper coffee-pots.

A coffee-pot and stirrer

Any stranger who calls is sure of a welcome and a drink of coffee. No visitor, even if he belongs to an enemy tribe, will be harmed while he is a guest.

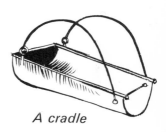

A cradle

During the day, Nasir's mother has many jobs to do. She spins sheep's wool and camel hair into long threads. With the threads she makes clothes and repairs the tent. She milks the camels and goats, and makes butter-milk, called *leben*.

The Bedouin Arabs have very simple food. Usually they live only on camel's milk, cheese and a few dates. For a feast they roast locusts, or kill and cook a camel or a sheep. They also eat boiled rice and leben.

All the men sit cross-legged on the floor and help themselves from the great dishes.

Nasir uses his right hand to eat. He picks up some food, rolls it into a ball and puts it in his mouth.

After the meal, Nasir and the other boys play a game with tent pegs, or they throw stones with their slings.

In summer the grass dries and even the winds are terribly hot. Often the wind blows so hard that it makes a sandstorm, and clouds of sand blow everywhere. Nasir curls up in his cloak. The camels lie down and close their nostrils.

When the storm is over, there is sand everywhere, in Nasir's clothes, in his mouth and eyes, and even in his food.

In the summer it is far too hot for Nasir to stay in the desert, so he and his family camp near an oasis. Here there is a well with water for the animals. The people who live at the oasis have a few date and pomegranate trees, and small fields of maize.

They draw water from the wells to water the fields. They live in houses made of baked mud, with flat roofs and small doors and windows.

In parts of the desert, oil wells have been drilled. This has made some Arab chiefs very rich. They have splendid palaces, and they have built hospitals and schools for their people.

How to draw plans

Here is a picture of a cup and saucer.

Here is a plan of them. The plan shows how they look from above. Can you see the handle of the cup?

Try to draw a plan of this jug.

Here is a plan of a table laid for breakfast.

How many cups and saucers are there?

Where is the teapot?

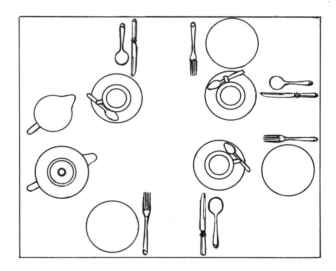

Here is a picture of a desk and chair.

Here are the desk and chair as they look from above. Can you see the book and the inkwell?

Here is Ann's house. It has no upstairs, and it is called a bungalow.

Here is a plan of the bungalow. The builder used a plan like this when he built the house.
How many bedrooms are there? How many rooms have doors which open on to the hall?

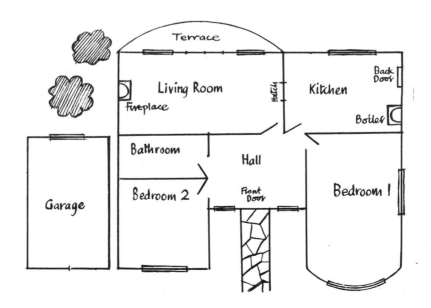

Here is a picture of part of a classroom. The plan below shows the whole classroom. Look at the plan. Part of it is drawn darker than the rest, and it is cut off by dotted lines. That is the part which is shown in the picture.

Key

B Boy
DB Draining Board
G Girl
R Radiator
S Sink
T Teacher

The classroom is 11 metres long and 8 metres wide.

How many boys and how many girls are there? Where is the teacher standing?

Draw a plan of your classroom, and show which is your desk.

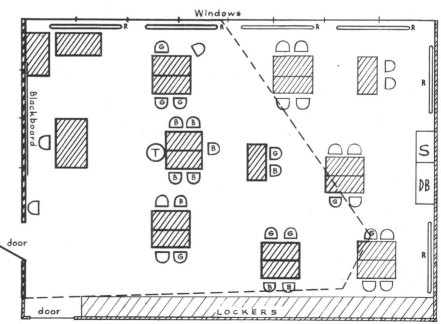

David of
Great Britain

David is the son of a farmer. He lives on a farm in the north of England. Because his father keeps animals, and grows crops as well, the farm is called a "mixed" farm.

David's father keeps cows, pigs, sheep and hens, and he grows wheat and "roots", such as turnips and swedes.

"Roots" are used to feed the cows and sheep in winter. David's father also grows potatoes to sell.

A turnip and a carrot

Here is David's farm seen from the hills nearby. Behind the house is a yard, and the farm buildings are built round the yard.

Not far from the farm is the village. Can you see the tower of the church?

34

Here is David's farm again. Find out what happens in each part of the farm, by looking at the plan below.

Key

A Farm-house
B Stockman's House
C Toolshed
D Granary
E Stable and Calf House
F Tractor Shed
G Dairy
H Cowshed
J Straw Yard
K Hay Barn
L Pigsty
M Store Shed
N Stack Yard
P Pond

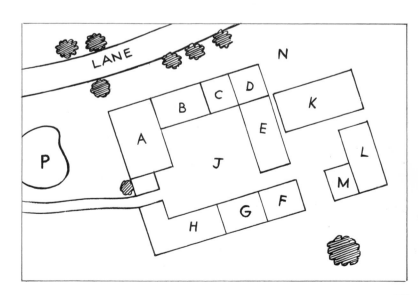

There is always plenty of work on a farm, on every single day of the year.

Each morning at about six o'clock, and again at four o'clock in the afternoon, the cows are milked. After the milking the cowman cleans the milking machine, the cooler and the churns.

Then he washes out the milking shed, and the churns of milk are taken by a lorry to the town.

David has his breakfast at seven o'clock. He has porridge, eggs and bacon, for there is always plenty to eat on a farm.

One of David's jobs during the holidays is to collect the eggs. He also helps to feed the pigs, the horse and the hens.

In the old days there were many horses on the farm. But now David's father has a tractor to pull his machines and carts, and he can do his work much more quickly.

Let us spend a day with David in each season of the year.

Spring

In spring the fields and hedges become green. The lambs which were born in January are now much stronger.

David drives the cows to the meadow. They have spent the coldest part of the winter in the farmyard and they enjoy eating the fresh green grass.

Later David watches the potatoes being sown by machine. After a few weeks dark green leaves are sprouting through the soil.

This machine digs trenches, plants the seed potatoes at regular intervals and then covers them up.

This machine only covers up the seed potatoes. Another machine must first dig the furrows and plant the potatoes.

Summer

Haymaking is one of the most important jobs in summer. The grass is cut on a fine day, when it is about 60 cm tall. A mowing machine cuts it close to the ground. The grass is left for several days to dry. When it is quite dry it is called hay.

A mowing machine

Then a pick-up baler moves round the field. It gathers up the hay and packs it into bales. The bales are tied with twine and dropped back on to the field. Later the bales are collected, stacked, and covered with a plastic sheet to keep them dry.

Sometimes the hay is stored in a dutch barn

Then in the winter, when the grass does not grow, the cows have plenty of hay to eat. David's father also gives them cattle cake, made from the seeds of cotton plants.

A haystack

A plough

Autumn

Wheat

In the autumn the wheat is harvested. Later on, next year's wheat is sown. The plough turns over the soil, and the harrow breaks it up. Then the tractor pulls the seed drill, and the seed goes down the spouts and into the soil.

Ten days later the first green shoots appear. They do not grow very much during the winter. But when spring comes again, the wheat is strong, and has plenty of time to grow and to ripen.

A harrow

In another field the tractor and spinner spin out the potatoes. David goes down the rows picking up the potatoes and popping them into his sack.

A cart collects the sacks, and then David's father sends the potatoes to be sold at the market.

Spreading manure

"Laying" a hedge

Winter

There are no idle days on a farm, even in winter. When the ground is frozen hard, that is the best time to spread manure on the fields. The hedges are trimmed in the winter, and every few years they are "laid", which makes them very thick and strong. Before the wet weather comes, the ditches are cleared out so that heavy rain can run off the land and flow away easily.

Feeding a lamb from a bottle

Winter is the time when the lambs are born. The shepherd builds a lambing pen with hurdles and straw, to protect the lambs from the wind. He covers the ground with straw so that the lambs will be warm. The shepherd lives in a hut near the pen. The mother sheep are called *ewes*. They are fed on turnips and hay. The lambs take milk from the ewes, but if a ewe dies the shepherd feeds her lamb from a bottle.

David enjoys going to the market with his father. Early in the morning they load the young cows, called heifers, into the cattle truck, and take them into town.

The heifers are driven into the ring and sold by auction to the farmer who offers most money.

Near the ring are the pens where the pigs and sheep are sold.

The town is very crowded on market day. People come in by bus from all the nearby villages.

41

At the market David's father likes to look at the new machines. Here he is looking at a small caterpillar tractor.

He does not need the big machines often, so he hires them in the town.

In one part of the market, farm produce is sold. David's mother brings in eggs and chickens to sell. The market sells everything: cloth, buckets, pans, clothes, vegetables and fruit.

Chai of China

In a small village in China lives a little boy called Chai. Chai's winter clothes are padded, so he looks as if he is dressed in a suit made from a quilt. In this part of China most of the trees were cut down many years ago, so there is hardly any wood to build houses. Chai's house has a bamboo frame.

Bamboo is a tall, strong plant which is used for making rafts and sails, houses, tiles and ropes.

The bamboo is tied together to make the walls, then it is covered with mud. When this has been baked by the sun it makes a good wall. A tile roof keeps out the rain.

There are only three rooms in the house and they are very crowded.

Outside in the courtyard Chai's father sifts the rice which he grows. His water-buffalo, his pigs and hens and his donkey live in the yard.

Chai's sisters keep silkworms on trays in the living-room. Each worm eats twice its own weight of mulberry leaves every day. The silkworm spins a fine thread round and round itself until it has made a *cocoon*.

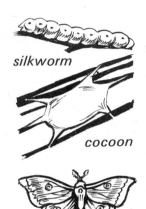

silkworm

cocoon

moth

Then Chai's mother kills the worm in boiling water, and unwinds the silk thread. (If she did not kill the worm it would change into a moth and break the threads.) From each cocoon she gets nearly *a kilometre* of silk thread. She sells the silk to a merchant in the town.

Chai's mother also prepares the dinner, which is nearly always rice. The rice is boiled in a big iron pot, with beans and bamboo shoots. Sometimes Chai also has fish, or an egg.

Chai sits on the floor to eat his food from a bowl. He does not have a knife and fork, but uses chopsticks, which are short sticks made of bamboo.

Chai's father has a few small fields near the river, where he grows rice. To grow well, rice needs water round its roots, so there are ditches full of water by all the fields. Chai's father pedals at a treadmill to lift water from the river into the ditches.

Chai helps his father to sow the rice in the seed beds. When the seeds begin to grow, a hole is made in the wall of the ditch so that water runs all over the beds. Then the hot sun and rich soil help the seeds to grow quickly.

Next, their water-buffalo pulls a heavy plough across another flooded field; this makes it ready for the little seedlings which are now about twenty centimetres high.

Chai and his family pull the seedlings out of the seed bed and plant them out in the flooded field. They push them into the mud under the water, so that only their tops are showing.

45

Cutting rice

If the rice fields are very dry, Chai's father lets water from the ditch run on to the fields. If there has been rain he drains some water away, so that the top of the rice is out of the water. When the rice begins to ripen he drains all the water away. Chai uses a sickle to cut the rice.

Threshing rice

Then his father threshes the rice by beating it against a big box. The grain falls to the bottom of the box. The straw is saved for the roofs of houses, and for making hats.

Rice

The rice does not take very long to grow, and Chai's father can grow two crops every summer. In the winter he grows wheat, beans and barley.

Sometimes Chai's brothers carry a live pig to market on a pole.

Chai's father has only his family to help him on his small farm. Some farmers nearby have joined all their farms together, to make one large farm. They have bought a tractor, but Chai's father cannot afford one for his farm.

Chai likes to walk down the muddy road to the Yangtse river, one of the biggest and busiest rivers in the world. The river is covered with flat-bottomed boats, called sampans.

Many people live all their lives in these boats, and they hardly ever set foot on land.

They keep birds called cormorants which catch fish for them. The birds have rings on their necks, so that they cannot swallow the fish which they catch.

The young children who live on the boats have pieces of bamboo tied to their backs. If they fall in the river, they float!

Chai sometimes goes on the sampans, and one day he hopes to go on the big junks. These boats have square bamboo sails. They carry large cargoes of rice down the river and along the coast, to the great ports of China.

A junk

Sometimes Chai goes to the town with his father. In the old part of the town the streets are crowded. There are many carts and bicycle-rickshaws. The alleyways are narrow with shops on both sides. The shopkeepers put all their goods on benches outside their shops.

Water carriers carry pails of water on long poles, and one shop even sells boiling water.

In one alleyway Chai stops to buy a bowl of rice which he eats with chopsticks. Sometimes the rice is flavoured with vegetables or fish.

In the new part of the town there are fine new shops and flats. Chai sees modern buses, cars, and large factories.

One day China will be a rich country, for plenty of coal and iron have been found in the north. Factories are being built, as well as roads, railways and new houses. But Chai's country is so big that the work will take many years.

48

A Norwegian saeter (see page 51)

A cable car on the slopes above a Norwegian fiord

Sigrid of Norway

Sigrid is gliding over the snow on her way to school. The long thin pieces of wood on her feet are called skis.

Sigrid lives in Norway. Norway has so many mountains that there is little good land for farming. In winter it is very cold, and even in summer the tops of the mountains have snow on them.

In the mountains, the snow is packed so tightly that it becomes a solid river of ice, called a *glacier*.

The glacier moves very, very slowly down the valley until it melts and becomes a sparkling stream.

The stream runs on, tumbling over the rocks, until it falls down the mountain side as a waterfall.

A glacier

Ocean liners in a fiord

Sigrid's farmhouse is near the foot of a waterfall. The waterfall drives the electric motor which lights the house and drives the saw which cuts up the trees. Then the stream flows through the fields into a long narrow valley which has been filled by the sea. This valley is a *fiord*. There are many fiords on the coast of Norway.

49

Sigrid's house is built on a layer of stones so that it is always dry. The walls of the house are made of timber. The roof is made of turf and birch bark, and it slopes steeply so that in winter the snow slides off it.

The cows and goats spend the winter in the barns, and the lofts above the barns are full of hay. The horse has to climb up a slope to reach the loft.

In the storehouse nearby, oatmeal, potatoes, salted fish and cheese are kept.

One of Sigrid's jobs is to bring logs for the fire.

Her father made the heavy wooden furniture, and her mother wove the gaily coloured rugs.

Pine trees grow on the mountain sides. Higher up the mountain the soil is not deep enough for trees, and only grass will grow.

In winter the grass is covered with snow.

In summer these high pastures, called *saeters*, are fresh and green, and the cows and goats are taken there to graze. For the whole of the summer Sigrid and her brother and her mother live in a hut on the saeter. They look after the animals.

Each night and morning Sigrid helps to milk the cows and goats. The cans of milk, butter and cheese are sent to the farm on a long wire which runs down to the valley.

Then the butter and cheese are sent to the market town at the end of the fiord.

In the valley Sigrid's father cuts the long green grass. He dries it on poles, for it rains too often to leave it on the ground. He needs plenty of hay to feed his animals during the winter.

Drying fish outside   *Bergen, one of Norway's main fishing ports*

There are also many fishermen near Sigrid's home. They fish in the fiords, and in the open sea. Two or three boats go together, towing nets between them. They catch cod, herring and plaice.

Some fish are frozen and sent to England. Others are salted and packed in barrels for the Russians, or tinned or smoked for the Americans.

Some men near Sigrid's home are foresters. They cut down pine trees in the great forests on the mountain slopes. Then the logs are floated down the rivers to the saw mills.

The logs are cut into planks for building houses, or making furniture or making shallow fish-boxes. Paper is also made from wood, and the paper of this book is probably made from a tree which grew in Norway.

Ashok of India

The River Ganges is a great river. It flows from the Himalaya mountains down to the Indian Ocean. Millions of people live near the river.

One of these people is Ashok, a little Indian boy. Ashok's hair is black. His dark skin helps to protect him from the hot sun. All the year round it is much hotter in India than in Britain. But in the middle of the hot season there is heavy rain, called the *monsoon*.

Ashok's father wears a turban, a long strip of material wound round and round his head. His mother wears a dress called a *sari*. It is really a length of cotton material wrapped round her.

Ashok's mother does her washing by the river. She beats the clothes on the rocks to make them clean.

A woman dyeing clothes

53

Here is Ashok's house. His father built it on a frame made from bamboo which grows near his house. He plastered the walls and floor with thick mud, and then left them to be baked by the sun.

The roof is thatched with rice straw. It slopes steeply, so that the heavy rains run off quickly.

The house was easy to build, for it has no windows and only two rooms.

Outside the low doorway is a porch where Ashok sleeps on the hot summer nights.

Nine people live in the house, but Ashok does not think it is crowded, for the sun shines all the year round, and all the family are out of doors except at night.

Ashok's house has very little furniture: only a few wooden chests, a coconut oil lamp, some large jars of water and a few brass pots and pans.

Ashok eats rice at nearly every meal, and to make it more exciting he sometimes has fish or spices and pickles. Ashok takes food from the large bowl. He uses a big leaf as a plate, and puts the food in his mouth with his fingers.

It is dark inside the hut because the only light comes from the doorway. At night Ashok's mother lights a coconut oil lamp.

All the water in the village comes from the well. None of the houses has a tap. Ashok's mother collects the water in a jar, and carries it on her head.

55

Ashok's father has a few small fields scattered about the village. The soil is very poor, but if he works hard he can grow enough rice to feed his family, and still have a little left to sell.

When the rice is ripe, it is cut, and spread out in the yard behind the house. Then the oxen trample on it to loosen the grain.

Ashok's mother collects the rice grains in a basket. She throws the grain in the air so that the chaff and dust blow away. The grain falls back into the basket again.

Then she crushes the grain to get rid of the hard skin, and the rice is ready for cooking.

Millions of people live in India, and there is not enough food for them all. Some farmers do not grow as much rice as they could. Their fields are too small, and they never put manure on them. The Government is trying to teach the farmers how to grow more rice.

56

There are no proper roads in Ashok's village, and the temple is the only building made of stone. On his way to school Ashok sees the potter making water jars and pots from clay. The potter bakes them in his oven. Then he spreads them in front of his shop.

At school Ashok often has lessons in the open air, in the shade of a banyan tree. Not every child in India goes to school yet, and Ashok will not be able to stay at school for long, for he must help his father in the fields.

Ashok enjoys his visits to the bazaar in the town nearby.
Twice a week traders lay out their goods in the narrow
alleys off the main street. The fruitseller sells many
brightly coloured fruits—bananas, oranges and mangoes.
Flies swarm over the sweets on the sweet stall, but
Ashok does not seem to mind.

The sacred white cows which wander up the street may
eat fruit and vegetables from the stalls. But nobody will
hurt them because they are holy animals.

In winter the bazaar is very muddy. In
summer it is hot and dusty. Ashok's
mother goes there to gossip and to learn
the latest news. She may buy spices to
flavour the rice, or cotton to make a
sari. But like all the people, she always
argues about the price before she buys
anything.

58

Piet of Holland

Piet is a Dutch boy. He lives in Holland, just across the North Sea from England. The journey from London takes about one and a half hours by 'plane.

Many years ago most of the land was under the sea. Gradually the hardworking Dutch people build walls of brick and stone, called *dykes*, to cut off the sea.

They used windmills to pump the sea water from the land. A windmill lifts the water into a canal, and more windmills and canals carry the water away to the sea. The new land, which used to be under water, is called a *polder*.

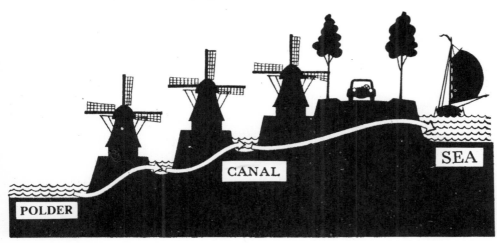

POLDER   CANAL   SEA

59

There are still many windmills in Holland, for in such a flat country there is nearly always enough wind blowing to drive the sails round. But nowadays most of the pumps are driven by steam or electricity.

Many of the new dykes are so wide that they have a road running along the top.

In some parts of Holland, where the land is below sea-level, the ground is soft and wet.

Builders drive dozens of *piles*, or posts, into the ground, and build their houses on them. They once used wooden piles but concrete is used now. Near Piet's house is a canal which drains the fields. In summer Piet swims in the canal, and in winter, when the canal is frozen, he skates on the ice.

Piet lives in a brick house. It has a red-tiled roof and wooden shutters. Trees shelter the house from the wind. There are not many woods in Holland because the land is so precious. But many roads are lined with trees.

60

Piet's father is a farmer. He owns four large fields which are very flat. In the autumn he plants out many bulbs— daffodils, hyacinths and tulips. In the spring the fields are full of colour, and the scent of the flowers carries for a great distance.

Daffodil   Tulip   Hyacinth

Picking tulips

People from all over Holland and even from other countries come to see the flowers. Piet sells huge bunches to the visitors.

Then the rest of the flowers are cut and thrown away, for only the bulbs are wanted. If the flowers were left growing they would take all the goodness out of the bulbs.

Piet helps to lift the new bulbs, and he splits them from the old ones. The new bulbs are packed and sent to other countries.

Not far from Piet's home there are many huge green-houses. Here tomatoes, grapes, lettuce and cauli-flowers are grown. They are sent to Britain, Belgium and Germany.

61

Piet of Holland

In those parts of Holland where there is good grass, the farmers keep many black-and-white cows, called Friesians. In the autumn the farmer puts blankets on their backs to keep them warm. But in the winter the cows live indoors in barns.

A Friesian cow

Each morning Piet sets off to school with his sister Bep. They find it easy to ride their bicycles along the straight road by the canal, because it is so flat.

Cycling is very safe too, because most of the roads have special tracks for cyclists. Nearly everyone in Holland has a bicycle.

Piet has already started to learn English at school, and Bep learns French and German as well.

In the summer the whole family goes on holiday on a sailing boat. In Holland there are many canals and lakes which are good for sailing.

Piet always takes his fishing rod on holiday, for there are many fish and eels in the canals.

62

Let's Remember

THE AMAZON FOREST

A hot, steamy forest

Peko's house is made from trees, creepers and leaves.

His father hunts animals and catches fish. His mother grows cassava.

GREENLAND

A cold, icy land

Kara lives in a wooden house which has two rooms.

Kara eats the flesh of seals. Her mother makes tents and clothes from seal skins.

SAUDI ARABIA

A hot, dry desert

Nasir's tent is easy to carry in the hot desert.

His family look for grass and water for their animals.

GREAT BRITAIN

Warm in summer

Cool in winter

Behind David's house are the barns, the dairy and the farmyard.

David's father keeps cows, sheep, pigs and hens. He grows wheat and "roots".

CHINA

Hot, wet summer
Cool in winter

Mud is plastered on a bamboo frame to make Chai's house.

Chai's father grows two crops of rice every summer. His sisters keep silkworms.

NORWAY

Warm in summer
Cold in winter

Sigrid's house is made from pine trees which grew on the mountain side.

Sigrid's cows spend the winter in the valley. In summer they live on the saeter.

INDIA

Hot, rainy summer
Warm in winter

On hot nights Ashok sleeps on the verandah of his house.

His father grows a little rice on his poor land. Oxen pull his plough.

HOLLAND

Warm summer
Cold winter

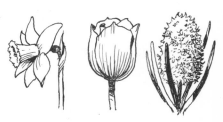

Piet's house is made of brick. Trees shelter it from the wind.

Piet's father grows bulbs on flat land which used to be under the sea.

Canals draining farm land in Holland

Dutch bulb fields

PART TWO

Looking at Everyday Things

The publishers are grateful to the following for information and for permission to reproduce photographs:

Aerofilms 105, 117a; Australian News and Information Bureau 126a and b, 127a and b, 128a and b, 129a and b, 130; Barnaby's Picture Library 69, 78, 86, 108, 123, 141c; British Petroleum Co Ltd 65, 139a and b, 141b; British Sugar Corporation 121a and b; British Trawlers' Federation 111, 113a and b; Central Office of Information 74, 77a, 110, 114, 120; Ceylon Tea Centre 135a and b, 136a and b, 137; The Clarke Collection of the Royal Meteorological Society 117b; Ernest Crowson Esq 124a, 125b; Danish Agricultural Producers' Information Service 106a and b, 107a and b; The Dunlop Company Ltd 75a and b, 76a and b, 77b; Fyffes Group Ltd 81a and b, 82a, b and c; Brooke-Bond-Liebig Ltd and Erich Hartmann-Magnum Photos 83, 84; ICI 67, 92a and b; Massey Ferguson (UK) Ltd 71; Meteorological Office 115, 116a; National Coal Board 100a and b, 102, 103a and b; National Film Board of Canada 70, 72a and b; Negretti and Zambra Ltd 116c; Popperfoto 80; Messrs Poulton and Freeman 133; Rank Hovis McDougall Ltd 73a and b; Royal Horticultural Society 124a, 125b; Shell International Petroleum Co Ltd 140a and b; Shell-Mex and BP Group 68; Tate and Lyle Ltd 118a and b, 119a, b and c; The Textile Council 87, 88, 89, 91; US Information Service 98

The diagrams on pages 101, 103 and 104 are reproduced by kind permission of the BBC Schools Department in Scotland, for whom they were prepared by the Isotype Institute. The weather forecast map on page 117 is reproduced by kind permission of *The Times*. The plan of the school on page 133 is reproduced by kind permission of the County Council of Essex, Messrs Poulton and Freeman, and H. Connolly Esq, CBE, FRIBA.

CONTENTS OF PART 2

Delivering oil to a house. The man is holding a radio, which stops the oil as soon as the tank is full

About part 2

It is Friday. Mr. Bell has gone to work, and Peter and Susan have gone to school. Mrs. Bell is clearing the breakfast table.

Oh! There is the doorbell ringing. The oil delivery man is at the back door. 'Six hundred litres?' he says. While he fills the tank, Mrs. Bell picks up the bottles of milk from the step.

She puts out a note telling the baker to leave two loaves.

Soon Mrs. Bell sets off for the shops. It is a cold day and rain is forecast, so she puts on her woollen coat and takes her nylon umbrella. Then she catches a bus to the town.

At the butcher's shop she chooses meat for Sunday's dinner. What shall she buy: Argentine beef or New Zealand lamb?

Shopping in a supermarket. Everyone has a wire basket in which to put the things they want. They pay at a desk on the way out

At the fishmonger's she buys a fillet of cod for her lunch.

The greengrocer has a window full of delicious fruit. Mrs. Bell buys bananas, apples and oranges. Then she buys vegetables. They are at the back of the shop where it is cooler.

Mrs. Bell goes to the supermarket and helps herself to tea, sugar, butter, bacon and eggs. She also buys matches, a frozen chicken and washing-up liquid. She pays for them on the way out, and catches the bus home.

Every day the Bell family needs many things from all over the world. This part of the book tells you about some of these things. It tells you how nylon is made, and how fish are caught. It tells you how coal is mined, and how sugar and cotton are grown. It tells you of many other everyday things, and about the way in which they come to our factories and shops.

In this part you will also read about hills and valleys, about winds and the weather, about rivers, maps, and the shape of the earth.

Wheat from Canada

Most of us eat bread at almost every meal, and we know that we can always go to the shop and buy a fresh loaf. Millions of loaves are eaten in Britain every day. Loaves are made from the seeds of a plant called wheat. Where does wheat come from?

Some countries, such as Britain, cannot grow enough wheat to make all the bread they eat. So they have to buy wheat from countries which have plenty of farmland, and fewer people.

Much of the wheat for the bread made in Britain comes from Canada. In Canada the winters are very cold, but the summers are hot and fairly dry, and the wheat grows easily. Because of the cold winter, most Canadian farmers sow their wheat in spring. It is called 'hard spring wheat'.

An ear of wheat

A prairie farm in Canada

70

Canadian children, playing on the ice

On page 70 you can see a farm in Canada. Round the farm buildings are trees and hedges, which give shelter from the wind. The wheat lands, called prairies, stretch for miles and miles, and the fields and farms are very large.

The winter is a quiet time on the farm. The land is covered by a thick blanket of snow for four or five months, and it is very cold. Many houses have double windows to keep out the cold.

When the children go outdoors they wear thick coats, and they often wear caps with flaps, to keep their ears warm. The older children play ice-hockey, which is very popular in Canada.

At the end of the winter the snow melts and the ground thaws. Then, in the spring, the soil is prepared for sowing the wheat seed. A harrow breaks up the soil, and the seed is sown in the warm earth.

Breaking the soil with a disc harrow

A combine harvester at work on the wide plains of Canada

The wheat grows and ripens during the hot summer. When it is golden brown, and the grain is firm, the farmer cuts it. He starts cutting when the morning dew has dried, and sometimes works until midnight. As long as the wheat is dry it can be cut.

On the big level prairies, the farmer can make good use of his combine harvester. This is a wonderful machine which cuts the wheat and threshes it as well. The grain goes straight into a lorry, and the straw is blown back on to the field.

The lorry takes the grain to a country elevator. This is a storage tower where the grain is kept until a train takes it west to Vancouver, or east to the Great Lakes. Here the grain is stored in huge elevators until it is loaded into ships which take it all over the world.

Loading ships with wheat at an elevator

Some of the wheat comes to Britain. Britain cannot grow enough wheat to feed all her people, and in any case British wheat is too soft to make very good bread. So hard wheat from Canada, Australia or Argentina is mixed with the soft British wheat. Together they make good flour for bread-making.

At the miller's, steel rollers crush the wheat grain, and turn it into flour. The baker mixes salt, water and yeast with the flour to make dough. The dough is left in pans in a warm room. Then the yeast in the dough makes it rise, so that the bread will be light and full of air.

The dough will be taken from the mixing machine into the waiting pan

Machines fill the baker's tins with dough, and the tins are put in the electric oven. Two hours later the crisp, brown loaves are taken from the oven. Soon they will be in the shops. Remember next time you eat bread, that it may be made from wheat which grew in Canada.

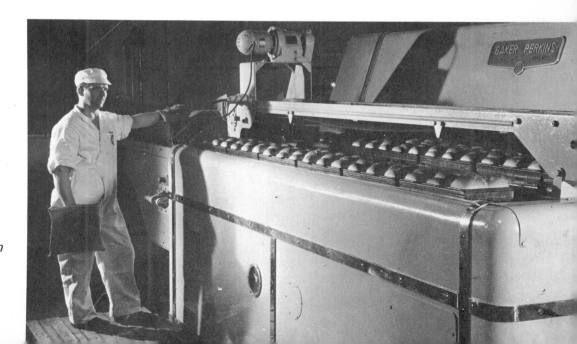

Taking the loaves out of the oven

Rubber from Malaysia

Tyres, gym shoes, balls, chair padding, garden hosepipes — all these things, and many others, are made of rubber. Where does rubber come from, and what is it?

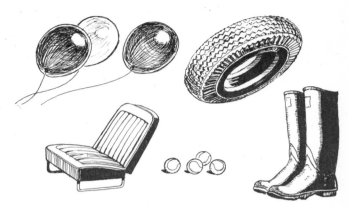

All these things are made of rubber

Rubber is made from a milky white juice, rather like the juice in a dandelion, which flows under the bark of the rubber tree. When the bark is cut, the juice, or *latex,* is collected in a cup.

Rubber trees grow wild in the thick forests near the River Amazon. There it is hot all the year round, and it rains nearly every day.

The rubber trees are scattered about the forest in little groups, a long way apart. They are hard to find. So about eighty years ago, rubber seeds were brought from South America to London.

Little trees were grown from the seeds and these trees were taken to countries with the same sort of hot, damp weather as the Amazon Forest.

A rubber plantation in Malaysia

74

Today rubber is grown in Indonesia, Malaysia, Thailand and Ceylon. In these countries there are plenty of men and women to look after the trees. In Malaysia many of these workers come from southern India and China.

Houses of estate workers

The workers earn more on the rubber estates than they could in their own villages in India or China, and they have better houses. Their houses are raised off the ground. They have overhanging roofs to give shade, and verandahs where the workers can rest in the evening.

It becomes dark by seven o'clock every night of the year in Malaysia, and it is never cold or wintry. Rain falls nearly every afternoon.

A rubber seedling

At first the rubber seeds are planted in baskets. As the seedlings grow, they are shaded from the hot sun. When they are strong enough, they are planted out, about four metres from each other, so that they have plenty of room to grow to full size.

Planting out a rubber seedling in its basket. Two seedlings are planted beside each planting peg. Later, the weaker seedling is removed

75

*The tapper makes a sloping cut
in the bark of the tree*

Small plants are grown near the seedlings. They keep the soil damp, by protecting it from the sun, and they help to keep down the weeds. The plants also stop the soil being washed away by the rain, or blown away by the wind.

After about five years the trees are nine metres high, and ready for 'tapping'.

The tapper, who collects the latex from the trees, starts work before the sun is hot, at six o'clock in the morning. He uses a sharp tool and makes a sloping cut, like a little shelf, half-way round the tree. He puts a spout in at the end of the cut, and on the spout he hangs a cup.

Very, very slowly the latex trickles into the cup.

Each tapper looks after four hundred trees, and taps half of them each day.

When the last tree has been tapped, the latex from the first is ready to be collected. Later, the latex is taken by a lorry to the factory.

Pouring latex into a lorry

76

At the factory, acid is added to the latex. This separates the rubber from the liquid. The rubber is rolled into flat sheets and hung on frames which are pulled into a heated shed filled with wood smoke. When the sheets are dry they are baled ready for shipment.

The rubber is rolled into flat sheets in machines like mangles

Pulling a rack of sheet rubber into the drying shed

Rubber is used most in countries which have many motor cars: in the United States of America, Britain and other countries in Europe. In these countries the rubber sheets are pounded, torn, kneaded and sliced in huge machines, to make them soft and sticky, like putty. Sulphur is added to harden the rubber so that it will bounce back into shape.

Rubber gloves are made by dipping *formers* into liquid latex. Tyres and hot water bottles are made in moulds from sheet rubber.

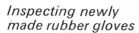

Inspecting newly made rubber gloves

But even the millions of rubber trees in Malaysia, Ceylon and Indonesia cannot produce enough rubber for our needs. So men have made *synthetic*, or artificial rubber. For some things it is better than natural rubber.

77

A mountain stream in Wales

The story of a river

It is a fine sight to see a stream in the mountains after a rainstorm. The water rushes along, splashing over the rocks. Sometimes it makes little waterfalls.

The stream gradually makes a narrow, steep-sided valley for itself. It flows so quickly that it carries stones and mud along with it. When the stones are first broken off by the stream, they are sharp and jagged. But the rushing water whirls them on, crashing them against each other and wearing them down until they are round and smooth.

At the end of winter, if the snow melts quickly, the stream becomes a torrent. In dry weather the stream is just a trickle, or it may even dry up.

Some of the rain which falls on the hills never reaches the stream, for it sinks into the ground. If it reaches a layer of solid rock, which it cannot pass through, it runs along the top of the underground rock until it finds a way out. Then it bubbles out of the ground as a spring.

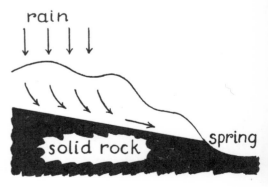
How a spring is formed

78

Here is a river which started as a mountain stream. Other streams, called *tributaries*, have joined the river making it larger. Now the valley is wide, and the hills on either side are gently curving. We can row boats and swim in this part of the river, for the water flows much more slowly than it did in the mountains.

No rocks stick out of the water, and the river is not flowing fast enough to move large stones. But it is still carrying along fine mud and pebbles.

At bends in the river the current wears away the outer bank. Near the inner bank, where the water moves more slowly, the sand and gravel sink to the river bed.

At bends in the river, the current wears away the outer bank

Sometimes flooding is a terrible disaster. This is an aerial view of part of Essex in 1953, when rivers overflowed and sea walls broke

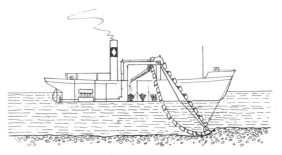

Dredgers have a chain of buckets which scoop up mud from the bottom of the channels

When the river reaches the flat land, it winds slowly to the sea, so slowly that it cannot carry even the finest mud, which it drops on the river bed.

When heavy snow has melted, the river cannot carry away all the water from the hills, and the flat fields nearby are flooded.

When the flood water has gone, these 'water-meadows' are covered by fine mud. This makes the grass grow well, so that there are heavy crops of hay, and rich grazing for cattle.

Some rivers have deep channels when they reach the sea, and ocean-going steamers can anchor in them. Docks are built where there is calm water so that ships can unload cargoes and passengers.

Sometimes there are mud banks at the mouth of a river. Then *dredgers* have to keep a channel clear. River pilots go on the big ships to steer them clear of the mud banks, which are marked by buoys.

How to grow bananas

Amos Jones is a negro who lives in Jamaica. He has saved some money and wants to grow bananas. How does he set about it?

First he buys a piece of land and clears away the tangled undergrowth. (Jamaica is a hot, wet country where trees, bushes and weeds grow easily.) Then Amos and his friend Len dig ditches to store the rain water, and put pieces of banana root into the ground.

Soon the roots send out shoots which grow into fine banana plants. After a year the plants are about five metres tall, with purple flowers.

Len is planting a banana root

When the flower petals fall, tiny bananas begin to grow. Each plant produces one bunch of bananas, and the bunch may contain over a hundred bananas.

When the bananas are fully grown, and while they are still green and unripe, Amos and Len cut them down. Amos uses a long pole with a knife at the end. He cuts into the plant, lowers the heavy bunch on to Len's back, and cuts off the bunch with his big knife, called a *machete*.

Watering the young banana plants

How to grow bananas

Soon Amos sells his first load of bananas. They are taken to a packing station. There, men cut off the 'hands' of bananas, and put them into cardboard boxes. Then the boxes are loaded into a ship.

In the holds (the store-rooms of the ship), the bananas are kept cool, so that they do not ripen on the voyage. The boxes are unloaded by special machines which carry them in canvas pockets so that they are not bruised. Then the bananas are hung in warm rooms where they ripen and turn from green to golden yellow.

Bunches of bananas grow upwards on the tree

A dealer buys the bananas. Soon Amos Jones' bananas are being sold in the greengrocers' shops.

The 'hands' of bananas are packed into cardboard boxes

At Southampton, the boxes are carried in canvas slings from the ship to the train

Meat from Argentina

Señor Pastor has an *estancia*, or farm, in Argentina. He keeps thousands of cattle. When the cattle have grown fat they are killed and sold for meat.

The flat grassland, where the cattle graze all the year round, is called *pampa*. It stretches for miles and miles. Although the soil is good, it is not damp enough for trees to grow.

A prize bull on Señor Pastor's estancia

Señor Pastor's farmhouse is large and up to date. Since his farm is far from the nearest town, he uses a windmill to make his own electricity.

Behind the house are workshops for the blacksmith, the harness maker and the men who look after the cars. There are garages for the cars and tractors, milking sheds and an office. Stables for the horses are also there.

Behind the estancia

Meat from Argentina

Manuel lassoing a young steer

Manuel is a cowboy on the farm. In Argentina he is called a *peón*. He has a strong horse and he is a very good rider. He rounds up the cattle by galloping round them, and he can lasso a steer even when he is riding at full speed.

Why do you think Manuel wears different clothes from a British farmer? Why does he have spurs on his boots, a hat (called a *sombrero*), and a neckerchief?

At least once a year, the cattle are rounded up for counting and branding. Manuel and the other *peónes* each lasso a steer.

To brand a steer, they make a special cut in its cheek. The cut shows that the steer belongs to Señor Pastor.

The steer is not branded on its body, as this would spoil its hide, which one day will be made into leather.

The cattle are driven into wooden compounds to be branded

But most of Manuel's days are not so exciting as you might think. For he has much hard work to do. He rides round the miles of wire fences, and mends any which are broken. He also has to see that the cattle are moved to a new pasture when they need fresh grass.

Manuel has to grease the windmills so that they turn easily. The windmills pump water from the deep wells to the troughs, where the cattle drink.

On part of the estancia, Señor Pastor grows *alfalfa*, which is like the clover we have in our fields. When the alfalfa is fully grown it is cut and left to dry. Then it is put into stacks rather like our haystacks.

Alfalfa

For most of the year there is plenty of grass for the cattle. But sometimes in the summer, when it is hot and there is no rain, the grass dries up. Then Manuel feeds the cattle with alfalfa.

When Señor Pastor has a few hundred good beef cattle which are fat enough for killing, Manuel and the other *peónes* round them up and drive them to the railway.

Meat from Argentina

Butchers wash and carve up the carcases

The railway trains take the cattle to a large factory at the mouth of the River Plate.

The factory is spotlessly clean inside, and all the workers wear white overalls. Here the cattle are killed, without pain.

Skilful butchers carve them up, so that every part of them can be used. Some of the meat is tinned and made into 'corned beef', but most of it is 'chilled', or partly frozen. It is not fully frozen, as that would spoil the taste. Special ships, with cold storage rooms, take the meat to other countries.

Most of Señor Pastor's meat is sent to Britain. From the docks it is at once taken to the big markets, such as Smithfield in London. Dealers buy the meat and sell it to the butcher, who soon has it for sale in his shop. He knows that beef from Argentina is good to eat.

This is Smithfield Market. At markets like this the butchers buy the meat they will sell in their shops. The market opens very early in the morning

86

Cotton from the USA

Let's look in Mr. Grey's shop windows. One is full of gay summer clothes. There are dresses, blouses, shirts and vests. Now look in the other window: there are materials, sheets and table-cloths.

Nearly everything in Mr. Grey's windows is made of cotton, which comes from the cotton plant. Cotton thread for sewing is made from the fluff inside its seedpod.

Cotton grows best in warm lands which have rain in the spring, followed by a warm, dry summer. Much raw cotton comes from the southern states of the U.S.A. Good cotton is also grown in Egypt, the Sudan and the West Indies. The cotton from India and China is not so strong.

In the U.S.A. the cotton planter has to be sure that all the winter frosts are over before he plants his cotton seeds. A few days after the seeds have been sown, the seedlings break through the ground. With the help of the rain and the warm sun, they grow quickly.

Planting the cotton seeds

Cotton from the U.S.A.

The seedlings are thinned, so that only the strong, healthy plants are left.

Weeds grow even quicker than the cotton plants, but hoeing between the rows keeps them down. Here you see a tractor hoeing three rows at once.

Cultivating the young cotton plants

By August, four months after the seeds were sown, the *cotton plants* are about one metre high.

Cotton is a relation of the hollyhock, as you can see from the shape of the creamy-coloured flowers.

When the petals fall, they leave behind the seed pods, called *bolls*.

Inside each boll are thirty or forty seeds, each the size of a little pea. When the seeds ripen, the boll bursts open.

Each seed looks rather like the seed of a dandelion, for it is covered with white fluff, called *lint*. The lint is the raw cotton.

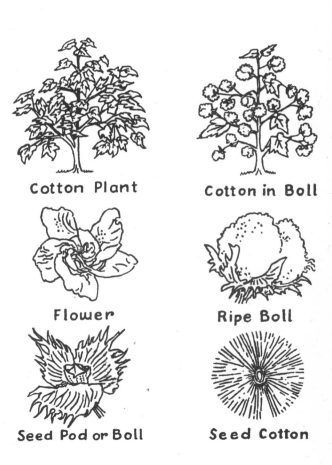

Cotton Plant   Cotton in Boll

Flower   Ripe Boll

Seed Pod or Boll   Seed Cotton

88

Although cotton grows easily, a heavy rainstorm can ruin the crop if the bolls have only just burst open. Sometimes a little insect called the boll weevil eats into the bolls. A low-flying aeroplane may be used to spray the fields to kill the weevil.

Whole families set to work picking the cotton when the bolls have burst. Not all the cotton is ready at the same time, and so the pickers take the lint from the ripe bolls only. Each picker puts the lint he has picked into a sack tied to his waist.

Machines can pick the cotton quickly, but not very carefully. They pick the leaves as well as the cotton, and the unripe bolls as well as the ripe ones.

This aeroplane is spraying the crops to kill boll weevil

Picking cotton by machine and by hand

Cotton from the U.S.A.

Most of the people who work in the cotton fields of the United States are negroes, whose ancestors were taken as slaves from Africa to America. Today there are no slaves in America, and many negroes grow their own cotton on their own land.

Cotton pickers outside their house

On the cotton plantations many of the workers live in wooden houses, with verandahs where they sit out on hot summer evenings. Every summer some of their friends from the city stay with them, and spend their holidays picking cotton.

The sacks of cotton are emptied into big baskets at the end of the rows. Carts take the cotton to a factory where a machine, called a *gin*, tears the fluffy lint from the seeds.

How a cotton gin works

1 *Seed Cotton*

4 *bars*

Air

3

Trash

6 *Lint Cotton*

Saw teeth

5

Seed

1 Seed cotton enters gin
2 Seed cotton is thrown against fast turning saws
3 Saw teeth take cotton up and against bars
4 Bars hold back seed but let lint cotton through
5 Seed falls down conveyor
6 Lint on saw teeth is carried away by a blast of air

Some of the seeds will be used for next year's planting, but most of them are crushed to squeeze out the oil which is in them. This oil is used to make margarine. The crushed seeds are made into cattle food, to feed cows in winter time.

This girl is working in a cotton spinning factory in Lancashire. She looks after six long machines like this one

The raw cotton is packed into bundles, called *bales*. The bales are wrapped in sacking, and held firm by metal bands. Then they are sent by ship all over the world.

Most of Britain's raw cotton arrives at Liverpool, where some is unloaded. The rest goes up the Ship Canal to be unloaded at Manchester, and sent to cotton mills in many parts of Lancashire.

In the cotton mills all the small threads of cotton are cleaned and combed by large machines. Then they are spun, or twisted together, so that they make one long thread. When the raw cotton is being twisted into thread, it must not get too dry, or the threads would break.

That is why most of our cotton mills are in Lancashire, where the air is usually damp. Looms weave the threads into fine cloths, which are printed with gay patterns and sold all over the world.

Cotton is an unusual plant. From it we make clothes, sheets and curtains, margarine, and even cattle food. Nylon is taking the place of cotton for some kinds of materials. But many people prefer cotton, because they like the 'feel' of it.

Nylon and rayon

Nylon does not grow on a plant like cotton, nor does it come from an animal as wool does. It is a new material, made by men and machines.

Chemicals made from coal, air and water are treated to make nylon *polymer*, which looks like chips of white marble. These chips are heated to make them into a liquid, which is forced through tiny holes.

As each jet of liquid dries, it becomes a strong thread.

Checking that none of the nylon threads is broken

The nylon is spun on to large reels called 'cakes'

Nylon is made into stockings and all kinds of clothes. Usually these clothes do not need ironing.

Sometimes nylon is mixed with cotton or wool to make the materials stronger.

Because nylon is strong, it makes good ropes, parachutes, fishing lines and firemen's hoses.

Rayon, or 'artificial silk', is another man–made material.

Let's build a house!

Would you like to live in a house like this? It has no gas, no electricity, no water and no drains. But there are other reasons why it would be a bad house to live in. Can you think of them?

Is this a good house?

A house in central Africa

A house in India

A house must give the right kind of shelter.

In the north of Canada, and in Greenland, houses must be specially made to keep out the cold. A house in the hot forests must be cool, and its roof must keep out the heavy rain. In Britain, and in other countries with the same sort of weather, houses must be warm in winter and cool in summer.

A house must be made of materials which are easily found nearby.

Most building materials are heavy, and it is costly to move them far. Houses in Sweden, and other countries with many trees, are often made of wood. Chinese houses have bamboo frames; Indian houses have roofs thatched with rice straw. Egyptians make houses of mud-bricks, dried in the hot sun. In Britain most houses are built of bricks, made from clay.

A house in Norway

A house in Britain

Let's build a house

Let's build a house! First we must decide where to build. Then a surveyor makes a careful plan of our plot of land. He shows the size and shape of the plot, and where are the nearest gas, water and electricity supplies, and drains.

Wood, stone, concrete or brick—which shall we use? Wood is expensive in Britain, because there are few forests. If we live in an area with a good local stone— near Aberdeen, in the Cotswolds, Derbyshire or parts of Yorkshire—we may use stone.

Concrete, on a steel frame, is very good for large buildings. But in most parts of Britain we shall probably use bricks, which can be made of local clay, or which can be brought from brickworks in other parts of the country.

The architect makes a plan of the house on his drawing board. The builders will work from this plan

An *architect* draws a plan of the house, showing every wall, door and window. The builder and his foreman mark out the shape of the house on the land.

The builder and his foreman mark out the land

94

Shallow trenches are dug along the lines which mark the position of the walls. Then the builder's men make concrete. Sand, cement, small stones and water are all churned together in a concrete mixer. When the mixture is poured into the trenches it sets as hard as a rock. This makes a firm base for the walls.

Making a cavity wall. Walls like this make the house warmer

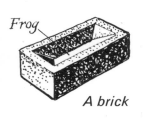

A trowel

Frog

A brick

Now the bricklayers start work. Here are the things they use. The trowel has a broad steel blade and a firm wooden handle. The bricks are 21·5 cm long, 6·5 cm high, and 10·25 cm wide. The 'frog' or hollow in the top helps to grip the mortar. Mortar, which is used with the bricks to make a wall, is made from cement or lime, mixed with sand and water. Like the concrete, it soon sets hard.

You can see that the wall is really two walls. The space between the walls prevents driving rain soaking through to the inside. It also helps to keep the house warm in winter and cool in summer. Can you think how it does this? Why does the bricklayer use a spirit level and a plumb line?

A plumb line

A spirit level

Let's build a house

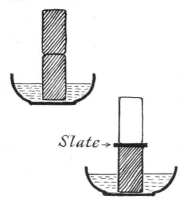

Slate→

This experiment shows how the damp course works

Just above the ground the bricklayer lays a damp course. This is a layer of bitumen or slate. Water cannot pass through it, and so the damp cannot rise up the wall.

To see how the damp course works, stand two bricks in water, one on top of the other. Water will slowly creep up both bricks. But if you put a slate between the bricks, water cannot pass through it, and the top brick will remain dry.

When the bricklayer makes a wall he *bonds* it. Can you see the difference between these two walls? The top one would easily crack. In fact, you could probably push it over yourself, because all the joins between the bricks are exactly above each other. The wall made in a *stretcher bond* is much stronger.

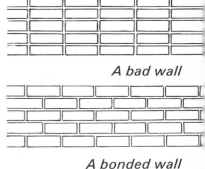

A bad wall

A bonded wall

Now the house is beginning to take shape. The bricklayers leave gaps in the walls, so that the joiners can fit doors and windows.

Ridge↘

Rafter

Eaves

The roof is built on a timber frame. First the *ridge* is set between the two ends of the roof. *Rafters* are fixed between the ridge and the *eaves*. Then *laths* are put across the rafters to take the tiles.

The roof tiles are made of clay. The nibs on the back of the tile are placed over the laths, and two nails are driven through holes in the tile to hold it firmly in place.

As soon as the roof is on, the builder's men start work inside the house.

The joiners lay the floors, and fit the doors, windows and stairs. The walls are plastered and the plumber fits water tanks, basins, sinks and lavatories. The electrician wires the house, and men from the gas, water and electricity boards connect the house to the main supplies.

To stop heat escaping through the roof the builder spreads a roll of fibreglass all over the loft. This insulates the house.

Last of all the painters set to work, painting and decorating the whole house, inside and out. As soon as the paint is dry, we can move into our new home.

The joiner

The plasterer

The plumber

The shape of the earth

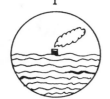

Our world is a huge ball spinning in space. For many years it was believed that the earth was flat, and long ago there were many sailors who would not sail far from land, because they were afraid of falling off the edge of the earth.

We can prove that the earth is round if we look through a telescope at a ship coming towards the land. This is what happens:

1 The smoke appears over the horizon.
2 We see the funnel and top parts of the ship.
3 The whole ship can be seen.

This drawing shows how the ship comes over the horizon. (The earth is not really curved as much as this, of course.)

This photograph of the earth was taken from a spaceship on its way to the moon. On the left it is day and on the right it is night. Near the middle of the picture you may be able to see the Mediterranean and the Red Sea.

98

The sun gives light and warmth to the earth.

At the *equator* (an imaginary line round the middle of the earth), the sun is almost overhead all through the year. The equator passes through the Amazon forest of Brazil, Kenya and the Congo. All these countries have hot weather every day.

At the equator

At the north and south *poles* it is very cold all the year round, because the sun is never very high in the sky.

At the poles

There are more temperate lands between the equator and the poles, lands where it is never very hot and never very cold. Britain is one of these lands; another is New Zealand, in the southern half (or hemisphere) of the world.

In temperate lands

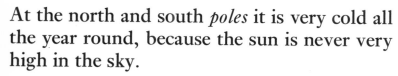

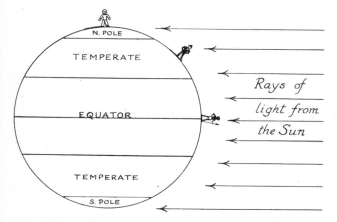

How the sun shines on the earth

At midday on March 21st a man at the north pole sees the sun just over the horizon.

In Britain the sun is halfway up the sky.

At the equator the sun is directly overhead.

The coal miner

Coal comes from under the ground. Many millions of years ago there were thick forests over our land. Slowly the forests were buried by mud and sand.

Deeper and deeper the forests were pressed under the earth. After millions of years the buried trees were changed into the hard, black stuff which we call coal.

A coal forest, millions of years ago

Here is a coal mine in Wales. Under the big tower is a hole, called a *shaft*. There are two of these shafts at the mine. The railway wagons stand by to take away the coal.

On the opposite page is a picture showing a slice cut through a coal mine. The picture shows the buildings at the pit top, the two shafts, and two *seams*, or layers, of coal.

A coal mine in Wales

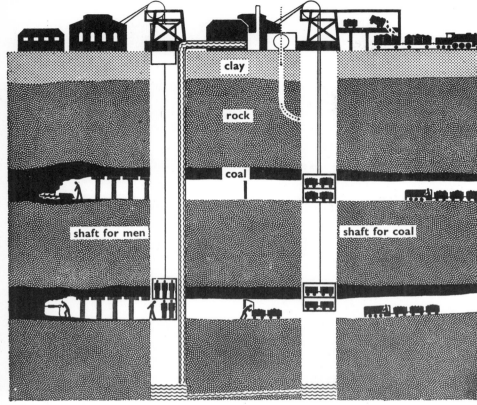

pithead baths   winding house   pumping house   ventilation fan   screening plant

A coal miner

clay

rock

coal

shaft for men

shaft for coal

This picture shows a slice through a coal mine

In this picture you can see:

pit head baths: the men have a bath here after work.

winding house: here is the engine which lowers the cages, and pulls them up again.

pumping house: the water which collects at the bottom of the mine is pumped to the surface.

ventilation fan: this sucks used air out of the mine, and draws fresh air in.

screening plant: the coal is washed, then sorted into sizes by the screens.

shaft for men: at the top is an empty cage; near the bottom is a cage full of men.

shaft for coal: near the bottom is a cage of empty tubs going down; half-way up is a cage of tubs full of coal. The shafts are wide enough for two cages.

A coal miner

Work in the mine goes on for twenty-four hours a day. There are three shifts, each of eight hours. Ted Bates is one of the miners on the morning shift this week. As he starts work, the men of the night shift are going home. When Ted finishes work, the men of the afternoon shift will take over.

When he arrives at the pit Ted changes into his working clothes. Why do you think he wears a helmet, boots with steel toe-caps, and knee-pads?

CAP LAMP

HAND LAMP

Collecting a cap lamp at the lamp room

Sometimes there are dangerous gases underground in the mine. These would explode if anyone lit a match. So Ted always leaves his matches in his locker before he goes down the mine.

At the lamp-room he collects his cap lamp and gives the lamp man a ticket, or check, for it. When there has been an accident underground, the manager looks at these checks, and quickly sees which men are still in the pit.

102

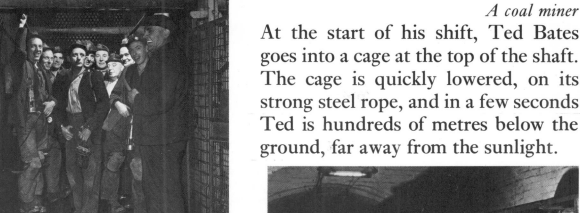

The cage

At the start of his shift, Ted Bates goes into a cage at the top of the shaft. The cage is quickly lowered, on its strong steel rope, and in a few seconds Ted is hundreds of metres below the ground, far away from the sunlight.

The place where Ted works, called the *coal face*, is two kilometres from the shaft bottom. He rides there on the *paddy mail*, a little train which carries the miners to their work.

The paddy mail

At the coal face Ted finds a heap of coal ready to be cleared. These three pictures show what the men of the night shift did, before Ted came to work.

Using a coal cutter to undercut the coal seam

Drilling shot holes in the coal face

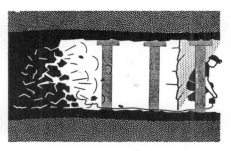

Blowing down the coal with explosive charges

A coal miner

Shovelling coal on to the
conveyor belt

The conveyor belt carries
the coal to the tubs

Hauling the tubs to the shaft

Ted Bates shovels the coal on to a
strong moving belt of rubber, called a
conveyor. This carries the coal to the
tubs. A battery locomotive pulls the
tubs to the shaft.

As he works, Ted puts timber or steel
props to hold up the roof. Later, more
permanent supports will be put in their
place.

Putting a wedge on top of
a steel prop

Coal makes gas, coke and
electricity for our factories and
homes

At the pithead the coal is washed, and
sorted into sizes by being shaken
through *screens* with holes in them.

Coal is used to make gas, coke and
electricity, as well as to heat our homes.
Even now that we have natural gas
from under the ground, oil and atomic
power, we still use a lot of coal.

A farm in Denmark

Food from Denmark

Every week we eat butter, bacon, eggs and cheese. These foods may come from farms in Britain, but it is more likely that they come from Australia, New Zealand, Canada, Holland or Denmark.

Denmark is a country which is famous for its farms. There the weather is a little hotter in summer, and a little colder in winter, than in Britain.

Most of the farms are small and the farmer needs only his family to help him. He keeps cows, pigs and hens, and to feed them in winter he grows root crops, oats and barley.

The cows graze in the fields during the summer. In winter they live in the cow house. Each farmer has only a few cows. Some are reddish-brown, called Red Danish cows, and others are black and white Jutlands, which give rich milk.

105

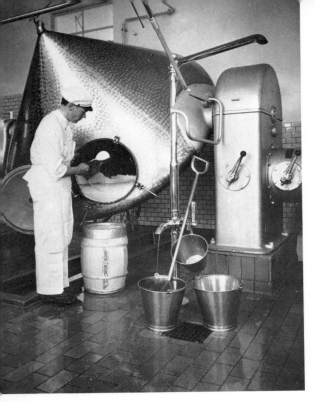

Making butter

Making cheese

At milking time the cows are driven into their stalls, or 'standings'. Each cow has fresh hay and water. The cows are milked by electric milking machines. Many farmers milk their cows three times a day. The milk is cooled and taken to the village dairy.

The dairy is owned by all the farmers. It is spotlessly clean and has fine modern machinery.

To make butter, the cream is separated from the milk, and then turned round and round in a large stainless steel churn. Some churns make a ton of butter at a time.

The butter is packed in foil, and most of it is sent to Britain with the label 'Lurpak'.

Rennet is added to some of the milk to make it go sour. Then the whey is drawn off, and the curd is made into cheese. The vat on the left holds about 5,000 litres of milk. The milk will be made into Danish blue cheese.

There are many other varieties of Danish cheese—the Danes even eat cheese for breakfast. But to the farmers, their pigs are even more important than butter and cheese.

106

Danish school children

Fishermen unloading their catch at Esbjerg in Denmark

The farmers feed their pigs with skimmed milk. When the pigs are heavy enough, they are killed for bacon. The bacon is 'cured' by being put into a brine bath. (Brine is salt water.)

The farmers' wives usually look after the hens, and feed them with grain grown on the farm. They send the hens' eggs to packing stations. There the eggs are tested by 'candling'—holding them over a strong light to see that they are good. Then they are graded in sizes, stamped and packed.

Many Danish chickens are reared for killing and eating. They are plucked and prepared at poultry 'dressing' stations.

Often there are bedding factories near these stations—can you think why?

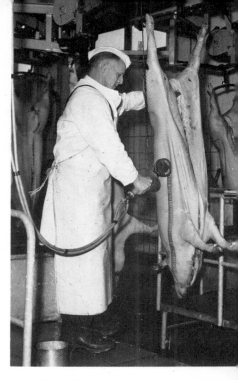

Printing 'Danish' on Danish bacon

An egg packing station

Danish farmers produce far more eggs, bacon, butter and cheese than the Danes can eat. Some of this food goes to Germany, France and Italy, but most of it goes to Britain. In exchange Britain sells coal and iron, tractors and machines, to Denmark.

A rough sea at Scarborough

A land yacht

Winds and directions

This trolley needs no one to push it. The wind fills its sail and drives it quickly along.

Trees near the sea often have strange shapes because on most days the wind blows in from the sea. A strong wind makes huge waves which crash over the sea wall.

How fast is the wind blowing today? Here is a table which will help you to decide.

Speed in km/h	Name of wind	What we see and hear
0 km/h	calm	smoke rises straight up
8 km/h	light breeze	leaves rustle
32 km/h	fresh breeze	small trees sway
48 km/h	strong wind	telephone wires whistle
80 km/h	gale	chimney pots and slates blown off
96 km/h	strong gale	trees uprooted and houses damaged

The direction of the wind helps us to know what sort of weather to expect. In Britain in the winter, winds from the south and west are warmer than those from the north and east which have been blowing across icy lands and seas.

A weather vane

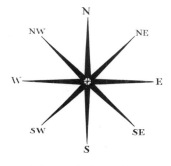

The points of the compass

Look for weather vanes on the tops of high buildings. The arms of the weather vane point north, south, east and west. An arrow above them swings round, and points to where the wind is coming *from*. The weather vane above shows that a west wind is blowing.

But pilots, sailors and many other people need to know more about wind directions than just north, south, east and west. The direction halfway between south and east is described as south-east. The diagram above shows eight points of the compass. What is the direction half-way between north and west?

Draw a wind-rose to show which way the wind is blowing. Each day look at the weather vane, and fill in one box to show where the wind is coming *from*.

This wind-rose shows that the wind has blown for two days from the west, for three days from the south-west, and for two days from the south.

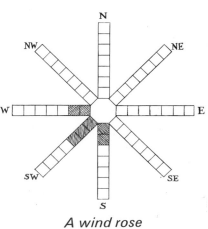

A wind rose

A trawler goes to sea

Here is the *Red Queen* making headway in a rough sea. Every now and then a big wave swamps her decks, and the little ship rolls and pitches. She is a trawler from Aberdeen, and now she is far out in the North Sea.

On the mess deck the crew are having a meal. When the ship rolls, the plates cannot slide off the table because they are held by bars of wood. The table is screwed to the floor, and cannot move.

The crew sleep in bunks, built one on top of another into the bulkhead, or wall. The bunks have sides, so that the men sleeping in the bunks are not thrown on to the floor in rough weather.

When the ship is out fishing the crew may spend as much as eighteen hours a day on deck. They work in the cold wind and salt spray, catching fish for us to eat.

The crew eat their dinner in two shifts

110

The *Red Queen* is thirty metres long, and six metres wide. Here you can see the different parts of the ship. There is plenty of space for the fish.

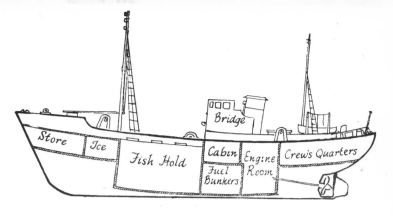
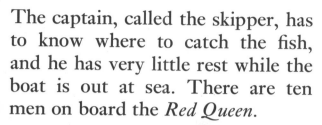

The parts of a trawler

The captain, called the skipper, has to know where to catch the fish, and he has very little rest while the boat is out at sea. There are ten men on board the *Red Queen*.

Mending the nets

The mate and a second fisherman see that the nets are always in good order, and that the deckhands pack the fish well. The chief engineer looks after the ship's diesel engines. A second engineer, and a 'trimmer', help him. Three deckhands put out the net, take it in, and mend it. They also gut and wash the fish which are caught in the net.

The cook works hard to prepare enormous meals for the hungry crew.

The crew have to gut the fish as soon as it is caught. If they did not, the fish would go bad before reaching port

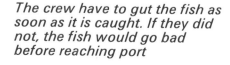

111

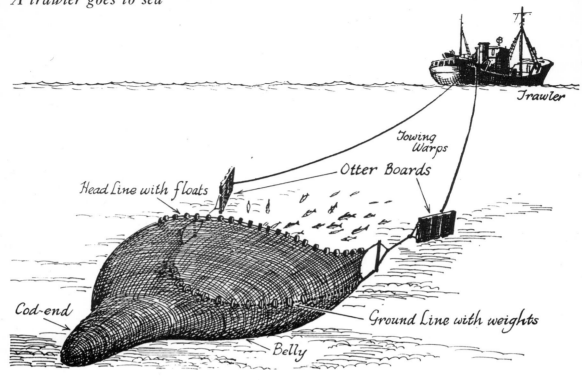

A trawler pulls its net along the sea bed

Haddock

Plaice

Cod

When the *Red Queen* reaches the fishing ground the skipper tells the crew to throw out the trawl. This is a cone-shaped net which is pulled along behind the ship. The bottom of the net is kept down by weights, the top is held up by floats, and the mouth is kept open by otter boards.

For three hours the *Red Queen* pulls the trawl along the bed of the sea. It is a heavy load even for her strong engines. Fish which feed on the bottom of the sea are disturbed by the ground rope. They swim up and pass into the net. Once the fish are in the cod end they cannot escape.

While the boat is trawling, the skipper watches the echo sounder. This sends sound waves to the sea bed. Any shoals of fish below the boat show as dark patches on the screen of the echo sounder. When the skipper thinks the trawl must be full, he orders it to be pulled in. The winch slowly winds in the wire rope.

All the men are excited now, for as the cod end comes up they can see the fish, flapping and struggling. A rope is put round the net, and the cod end is lifted from the water and held over the deck.

The skipper watches the echo sounder

Halibut

Hake

Sole

The mate pulls a knot and all the fish fall on to the deck—sole, plaice, cod, haddock, halibut, whiting and turbot. The mate at once ties up the cod end, the trawl is put over the side, and trawling begins again. In one trip the trawl may be thrown out and hauled in fifty times.

Putting out the trawl for the second catch

113

The fish are gutted, so that they do not go bad. Then they are washed, sorted, and stowed in the fish hold.

They are laid on shelves lined with ice. The ice keeps them fresh while the *Red Queen* is racing back to port.

At the harbour the fish are quickly unloaded and sent to market where the whole catch is sold. Then it is sent by train to all parts of the country, and soon it is on sale in the shops.

The *Red Queen* takes on ice, oil and stores of food, and within thirty-six hours she is ready for her next trip.

Billingsgate Market is very busy early in the morning. It handles all the fish for London

114

'Here is the weather forecast'

Red sky at night,
Shepherd's delight.
Red sky at morning,
Shepherd's warning.

Verses like this were very useful before the B.B.C. began to give us a weather forecast. But now, several times a day, news about the weather is broadcast on radio and on television.

A weather ship

If there is a gale warning, ships can shelter in port. Aeroplanes can fly higher, above the clouds, to avoid bad weather. If frost is likely, market gardeners cover their plants.

The B.B.C. gets its news about the weather from the Meteorological Office, which makes the weather forecast. This office collects information from two hundred weather stations in the British Isles, and from 'weather ships' far out at sea. They send in reports four times a day. Aeroplanes, and balloons carrying instruments, are sent up to great heights to find out what the weather is like in the upper air.

These men are launching a weather balloon.
Special instruments will hang below it

Here is the weather forecast

The weather stations and ships have special instruments to tell them all about the air—how heavy it is, how much water it is carrying, how hot or cold it is, and how quickly it is moving.

The instruments are kept in a special kind of box, so that the sun cannot shine on them.

Reading the instruments at the London Weather Centre

A barometer

The *barometer* tells how heavy the air is.

To measure how hot or cold it is, that is to find the temperature, the weather stations use a *thermometer*. You may have a thermometer like this one in your classroom.

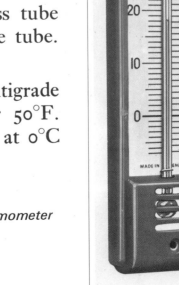

On warm days the red liquid in the glass tube *expands*, or grows bigger, and rises up the tube. On cold days it shrinks down the tube.

Temperature is measured in degrees, Centigrade or Fahrenheit, written like this: 10°C or 50°F. Water boils at 100°C or 212°F. It freezes at 0°C or 32°F.

A thermometer

The men at the weather station also look at the clouds. They notice what kind of clouds these are, how high they are, and which way they are moving.

Here are two kinds of clouds which they often see. The fluffy 'cotton-wool' clouds are called *cumulus*. The *cirrus* clouds, which look like a white horse's-tail, are often called 'mares' tails'. They are very high in the sky where it is so cold that they are made of tiny crystals of ice, not water.

Clouds help to tell us about the weather, but we can never make a forecast just by looking at them, so the men and women at the weather stations send reports to the Meteorological Office.

Here is the weather forecast

Cumulus clouds

Cirrus clouds—'mares' tails'. They are made of tiny crystals of ice

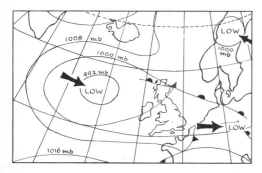

A weather map, from a newspaper

Wind, clouds, temperature, pressure —all these and many other details are put on a map of Europe and the Atlantic Ocean. The forecasters compare this map with maps made on earlier days. Then they say what they think the weather will be.

117

Where does sugar come from?

Do you like sweets—toffee, rock, bull's-eyes and peppermints? Most children do—and grown-ups, too—which is why there are so many different sweets to choose at the sweet shop. They are made with sugar, and sugar is used also for making cakes, puddings and jam.

Most of the sugar we use comes from the juice found in the stem of the sugar cane. This cane is a very strong grass, and it grows over three metres tall. The canes have long spear-shaped leaves, and feathery grey flowers.

Sugar cane will grow only in hot, wet countries such as India, Brazil and the West Indies. Much sugar is grown in Jamaica, one of the islands of the West Indies.

Sugar cane growing

Jamaica is a lovely island, with forests and mountains. But sometimes there are fierce hurricanes, and houses and trees are blown down.

A sugar factory in Jamaica

118

Planting a piece of cane

How is sugar grown? First, a short piece of cane, which looks like bamboo, is cut from an old plant. This piece of cane is buried in the ground. New shoots grow up from the joints, and soon a spiky leaf shows through the soil. The hot sun and the rain make the sugar cane grow quickly. Sometimes it will grow as much as 2·5 cm in a day.

The planter looks after the soil very carefully. Ditches run through his fields (you can see some of the women washing clothes in one of them). If there is too much rain, the planter drains the water away into the ditches. But if there is not enough, he lets in water on to the fields.

Drainage ditches

The cane grows for over a year, and when it is three to five metres high it is ready for cutting. Men use their sharp knives to cut through the cane close to the ground. Then they chop off the green top, and any other leaves that remain.

More and more of the cane is now being cut by harvesting machines like this one

119

It is not easy to cut the canes by machinery because they are so tangled. But on many estates, machines with a grab, like huge hands, tear out the cane.

When all the cane is cut, it is loaded into carts and taken by railway to a factory. Here a machine chops up the canes.

Then the canes are crushed between enormous rollers so that every drop of sugar juice is squeezed out.

Some of the juice, called molasses, is used to make thick black treacle. The rest of the juice is boiled in huge pans.

A crane lifts the sugar cane from a mule cart into railway wagons

As it cools, brown crystals form at the bottom of the pans. The crystals are put into sacks and sent to countries which cannot grow their own sugar cane.

When the sugar arrives in Britain it is taken in barges to the sugar refinery, where it is stored in silos. There are big sugar refineries near the ports of London, Liverpool and Greenock.

Working the controls of the boiling pans in a Jamaican sugar factory

Harvesting sugar beet in East Anglia

Most people in Britain prefer white sugar to brown. So at the refinery most of the sugar is boiled again. As it cools, crystals of sugar appear, sparkling and white. Some of this white sugar is cut into lumps, or made into icing sugar. Sugar can also be obtained from sugar beet, as well as from cane. Sugar beet is a beetroot with a fat white root. It grows well in cool countries, like Britain. Much sugar beet is grown in the east of England, in East Anglia.

When the beet are fully grown they are lifted by a machine, which also cuts off the tops. They are taken by road to a factory. There the beet are sliced and water is run through them many times to draw out the sugar, making a liquid.

The liquid is heated, and crystals of sugar are formed. The sugar is packed and sent to shops and food factories.

A sugar beet

Sugar from beet tastes exactly like sugar from the West Indies. When you buy sugar you cannot tell whether it has come from Jamaica or East Anglia.

121

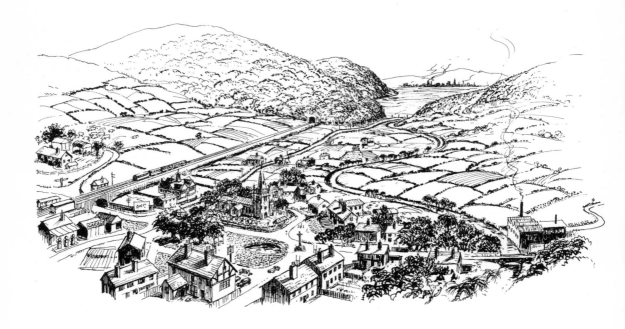

Valley and hill

Here is a valley, with a stream running through it. The village has a few houses, a church, a school, one small factory and a little railway station nearby.

Think about the picture, and see how many of these questions you can answer.

1 Where does the water in the stream come from?
2 Why is the village near the stream?
3 Why is the factory near the stream?
4 Why is the village in the valley, and not on top of a hill?
5 Beyond the railway is a farm. What crops will the farmer grow, and what animals will he keep?
6 Tom is leaving school and wants to work in the village. How many different jobs can you find for him?

Here is the moorland, high above the valley. Only sheep can live on the short grass, and they have to wander a long way to find enough to eat.

The grass is short, and hedges will not grow, because there is only a little poor soil over hard rock. Instead of hedges there are stone walls, made of stones which were quarried from the hillsides.

In places where the rain cannot soak through the rock, the ground is wet and boggy. In other places, the rain runs off the hillsides to form little streams.

There are few roads over these hills, only paths and sheep tracks. In winter the snow is often so deep that no one can cross the hills.

Growing vegetables

There are five members of the Banks family, and they eat a great many vegetables every week. But vegetables are often expensive in the shops, so Mr. Banks grows his own.

Mr. Banks sows his first row of peas in March. When the plants are a few centimetres high he puts in brushwood for them to climb up as they grow.

From June onwards the pods are full and ready to be picked. The pods have as many as nine peas in them, and each plant has many pods.

Putting up hazel stakes for runner beans

Green vegetables are good for the family, so Mr. Banks grows cabbages, cauliflowers and Brussels sprouts. He does not always grow them from seed. Sometimes he buys young plants from the market. Mr. Banks knows that he must water the plants if there is no rain.

Picking Brussels sprouts. The sprouts at the bottom of the stalk are ready first

124

Leeks, which taste rather like onions, are grown from seeds in frames. (Frames are boxes, covered with glass, which protect the young seedlings.) When the young leek seedlings are about 12 cm in height, Mr. Banks drops them into a hole made by a 'dibber'. The part of the leek which is in the hole does not get any light, and so it stays white. That is the best part to eat.

Mr. Banks sows his carrot seed in April and 'thins' the carrots in May. He thins them by pulling out the smallest carrots where they are crowded. While they are growing, Mr. Banks keeps them free from weeds, and he waters them in dry weather. By September the carrots can be lifted from the ground with a fork.

Mr. Banks grows many other things as well—lettuces, tomatoes, radishes, onions, parsnips, rhubarb and raspberries. He grows enough potatoes to last several months. All the family help him.

Planting out leeks with a 'dibber'

Planting radishes in a frame

125

The homestead
of a sheep farmer
in Australia

Wool from Australia

Mr. Driver is a sheep farmer in Australia. Here is his homestead. Nearby are the outhouses—the wool sheds, the garages for the cars and tractors, the stables, the smithy and the houses for the men who work on the farm.

The sheep station, as his farm is called, is very large. It has eight huge fields, called *paddocks*, each larger than a whole farm in England. In each paddock there are thousands of sheep.

The farm is over 150 kilometres from the nearest town. Many farmers have wireless transmitters so that they can talk to their friends who live in the district. If any one is very ill, Mr. Driver uses his wireless transmitter to send for a 'flying doctor'.

Carrying a stretcher to the flying doctor's plane

Here is one of Mr. Driver's *merino* sheep, which was Grand Champion at the Warrego show. It has fine crinkly wool which is very soft and can be made into the best quality woollen clothes.

A Grand Champion merino ram

The men on the farm have many different jobs to do. The boundary riders ride round the paddocks on horseback to see that there are no gaps in the wire-netting fences. The fences keep in the sheep and keep out the wild dogs, called *dingoes*, which would attack the sheep. They also keep out the rabbits which would eat the grass.

A dingo

Sometimes there is no rain for four or five months, and the grass withers in the hot sun. The sheep drink water from the troughs which stand in each paddock. But if the wells run dry and the water storage tanks are empty, then the sheep may die of thirst.

Both sheep and cattle drink at special troughs. The water is raised by windmills

127

In spring, when the weather begins to grow warmer, the stockmen round up the sheep, so that their fleeces can be cut. The stockmen have sheep-dogs to help them. The dogs drive the sheep into pens near the shearing shed.

The sheep are driven into pens. From the pens they go to the shearing sheds

The men who shear the sheep go from station to station. They are paid so much for each sheep they shear, so that the more sheep they shear, the more they earn. Some of the shearers are so quick that they can shear more than three hundred sheep in a day.

The shearers use electric shears, rather like the clippers a barber uses.

First they cut off the dirty wool under the sheep's body. Then the rest of the wool, called the fleece, is taken off, all in one piece.

The wool is sheared off in one piece

*Dipping the sheep
in a long trough*

After each sheep has
been shorn, it is *dipped*
in a ditch full of brown
disinfectant. This kills
all the insects which live
on its skin. The sheep
do not like being dip-
ped, for even their heads
are ducked under the
disinfectant.

Before the sheep go back to
the paddocks they are 'branded', so that if one of them
gets through the fence into
another farm the farmer will
know it is Mr. Driver's
sheep. His sheep are marked
with a 'D' for Driver in a
strong dye which will not
wash out in the rain. When
the sheep go back into the
paddocks they look very thin
but their wool soon begins
to grow again.

*Painting the farmer's mark
on the sheep*

Wool from Australia

At one end of the shearing shed the fleeces are 'skirted' by men who pull off all the dirty and tangled wool from the edges of the fleece. Then men called 'classers' sort the fleeces.

They class them by looking carefully at the wool, and noting its length, its colour, and how fine it is. Each class is placed in a separate bin. Then the fleeces are pressed, weighed and packed in bales. The bales are wrapped in sacking and taken by lorry to the railway station, many kilometres away.

Then the bales go by train to Sydney, or one of the other wool-selling centres, where the wool is sold by auction.

Ships take much Australian wool to Britain. It goes by train to Yorkshire and other places to be made into cloth.

Many sheep are kept in Britain but not nearly enough to make warm clothes for everybody.

That is why Britain buys wool from Australia, New Zealand, South Africa and Argentina.

Classers, sorting the wool

The bales are loaded on to a ship

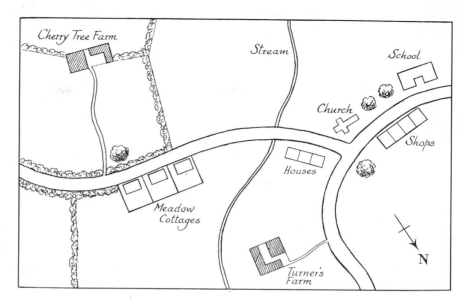

A map of
Crofton
village

Looking at maps

This is a map of part of Crofton village. Peter Todd lives at Cherry Tree Farm. When he goes to school he walks down the farm lane and when he reaches the road he turns left.

He calls for his friend Robert at Meadow Cottages. They cross the bridge over the stream, turn left at the church, and go into the school playground.

After school, Peter and Robert walk home. What do they see on their way?

Here is Peter on his bicycle. Will he turn right or left to go to the shops?

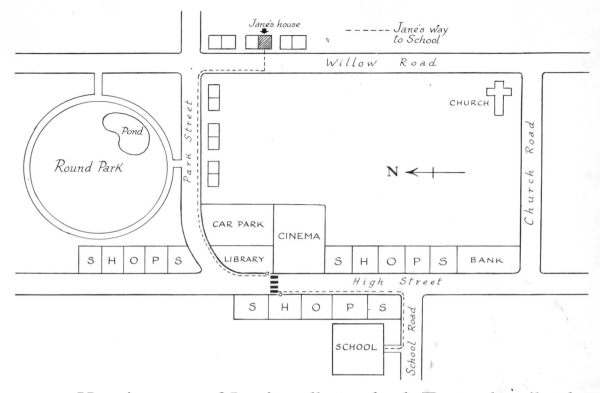

Here is a map of Jane's walk to school. Try to describe the way she goes. Begin: 'Jane crosses the road in front of her house, and turns right.'

Notice that she crosses the High Street at the Zebra crossing. Describe another way Jane could go to school.

Point to this place on the map. Is the cinema north or south of the bank? Is Jane's house east or west of the library?

Here is Jane's school. A path leads up to the main door, which is painted white. Just to the left of the door are the big windows of the hall.

The school has only six classrooms. You can see them on this plan.

One day Jane has to take a book from classroom 6 to classroom 3. Find out exactly which way she goes.

Jane finds a visitor in the school playground. Which way does she go to take him to the headmaster?

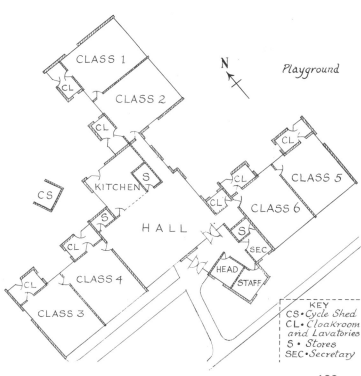

Tea from Ceylon

In Queen Anne's reign

These ladies of Queen Anne's reign are drinking 'tay' or tea. It cost forty shillings a pound, yet they were never sure of the best way to make it. Sometimes they boiled the tea, threw away the liquid, and ate the leaves. At other times they made their tea and kept it in barrels like beer.

In Queen Anne's days all our tea came from China. Today tea is also grown in India, and Ceylon, in Pakistan and East Africa. Nearly everyone in Britain drinks several cups of tea every day. If we shared out all the tea used in Britain in a year, every man, woman and child would have forty packets of tea.

A tea plantation in Ceylon. The women are picking the leaves from the tea bushes on the hillside

Here is a tea plantation in Ceylon. On the hillsides the flat-topped tea bushes are growing. Brown-skinned women are picking the leaves from the tea bushes. Nearby is the factory where the leaves are taken.

Planting tea cuttings

To make new tea plants, cuttings are taken from a big bush. The cuttings are put in plastic bags filled with soil. After being watered every day for a year, they are ready for planting out on the hillsides.

Tea plants like rain, but they do not grow well if water collects round their roots. They grow best on the hillsides, where the water drains away.

The tea plants grow all the year round: they are evergreens. If they were allowed to grow fully, they would be nearly ten metres tall. But they are pruned, so that the bushes are kept down to about one metre high, with flat tops.

Pruning a tea bush

While the bushes are growing, the soil between the rows is hoed regularly to kill the weeds.

The families which work on the tea estates live in small houses, built close together.

Because the sun is so hot, many of the houses in Ceylon have stone walls with very few windows. To provide shade they have overhanging roofs.

Houses of the workers on the tea estate

135

Plucking tea

Most of the workers on the tea estates are Tamils, from southern India. They earn far more in Ceylon than they could in their own villages in India.

When the tea bushes are about four years old, the fresh shoots are ready for plucking. In Ceylon, which is near the equator, plucking goes on all the year round. The pluckers are women. Their hands dart quickly over the bushes, nipping off the young leaves and buds.

They throw the tea into the baskets on their backs. As each basket is filled, the plucker takes it to the roadside to be weighed. A skilful plucker can harvest 25 kg of green leaf in one day.

The plucker takes two leaves and a bud

When the tea has been weighed, it is taken to the factory.

Inside the factory the tea leaves are spread on racks made of nylon. Here they are left to dry for a day. Then the leaves are put in a rolling mill, to release the juices in them. A drying machine turns them into the small, black tea leaves which we know. The leaves are sorted, according to size, by sieves.

Spreading the tea to dry on withering racks

The sizes have such names as Orange Pekoe, Fannings and Dust.

The tea is sold by auction. All the merchants bid for the tea, saying how much they will pay for it. The one who bids the most money buys the tea.

But before the merchant buys the tea, it has to be tasted. The taster makes a pot of tea from each kind of leaf. After six minutes he tastes some on a spoon. He says how it can best be mixed, or blended, with other teas.

Tea tasters in Ceylon

1 kg of green tea leaves, when dried, makes 250 g of tea

The blended tea is packed in plywood boxes, called tea chests. Each chest is lined with aluminium paper to keep out the damp, and holds about forty-five kilogrammes of tea.

Ships take the chests of tea to countries all over the world. When the chests arrive at the docks they are unloaded and stored in warehouses. Some of these chests are taken out each day, and the tea is put into the small packets which we see in the shops.

Lowering tea into a ship's hold

137

Oil from Saudi Arabia

What makes them go? All these things, which we see nearly every day, need oil to make them work. There are many kinds of oil: petrol is one kind, and paraffin is another. So it is oil which drives tractors, aeroplanes, diesel trains and cars. We use oil when we 'oil' our bicycles, and oil is used to make insect killers and paints.

How was oil formed? No one is certain. Do you remember how coal was made? Oil was probably made in a similar way. Millions of years ago many sea creatures died, sank, and were buried by mud. The sea creatures were pressed together until the mud became rock, and the creatures became thick, greeny-black oil.

Another name for oil is *petroleum* which means 'oil from rock'. The layers of rock hold the oil, rather like a sponge holding water. Oil is often trapped by *impervious* rock above it, and cannot escape, even though there may be gas and water with the oil which are trying to force the oil out.

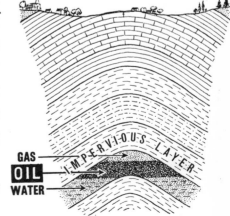

Underground rock, filled with oil

GAS
OIL
WATER

How is oil found? In the early days of the oil industry, about a hundred years ago, many wells were drilled in places which could never produce oil. Nowadays a very careful search is made before drilling begins. Men called geologists try to find places where there may be oil.

A geologist leaving a survey helicopter

Sometimes helicopters are used in the search for oil. They carry instruments which show which rocks may contain oil.

Geologists chip samples from the rocks, and try to find out whether there *could* be oil underground. But the only way they can be certain that there is oil, is to drill for it.

Reading a geological instrument

Before an oil well can be drilled there is often a great deal of work to be done, preparing the site. Nearly always a road must be made. Often a swamp has to be drained. When this has been done, a derrick is built. This is a tower, about 50 metres high, which holds the top of the drill pipe. The bit—the part which bites into the ground—is made of very hard steel.

An oil derrick

Lowering the drilling bit

The drill pipe is turned by an engine. As it goes round and round the teeth of the bit cut through the rock, and the hole becomes deeper and deeper. Some wells are more than 5 km deep. When the oil is reached, the gas and water try to force their way out, so pushing oil up to the surface. In some wells the oil has to be pumped out.

From the well the crude oil is pumped into storage tanks. Then it is pumped through pipelines to the ports.

A children's hospital, built by an oil company in Brunei

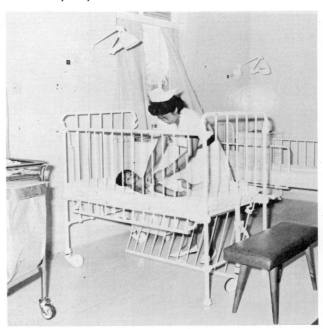

Oil is used all over the world. Most of the world's oil comes from the U.S.A., Venezuela, Russia, and the Middle East. Oil is often found in very poor countries — round the Persian Gulf, for example. In these countries the oil companies build houses, hospitals and schools.

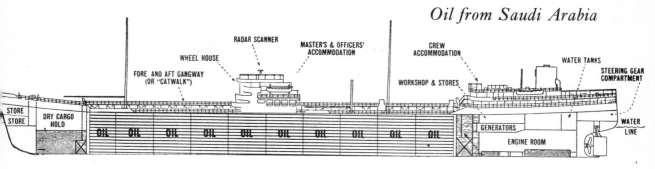

The parts of an oil tanker

Special ships, called oil tankers, bring the oil to this country. They have many tanks to hold the oil.

When the tanker reaches port the oil is pumped into storage tanks. But this crude oil cannot be used as it is, and it is taken to a *refinery*. There it is split up into the oils we use—paraffin, diesel oil, bitumen, petrol and hundreds of others. There are big refineries at Shell Haven on the Thames Estuary, Stanlow on the River Mersey, and Fawley near Southampton.

Road tankers carry the different oils from the refinery to our factories, garages and shops.

Part of an oil refinery

Refuelling an airliner

141

Do you remember?

Susan has come home from school, and is having her tea. Mrs. Bell is getting supper ready for the rest of the family who come home later. In the kitchen you can see some of the food which Mrs. Bell has bought this week.

Make a list of all the different foods you can see. (There are more than twenty altogether.) Some of them come from our country, others come from countries in various parts of the world. Against each food, write down the names of the countries it comes from, like this:

Butter comes from Denmark, New Zealand and . . .

Suppose you are a man who grows or makes one of these foods, or a fisherman. Describe a day in your life.

Here is a corner of the High Street in a small town. There are people at work, and things to buy in the shops.

Write down a list of all the useful materials you can see, and say where each comes from, like this:

Rubber comes from Malaysia, Indonesia and . . .

Now make a list of everyday things which are made from rubber, and from all the other materials in your list.

Some of the things in the picture, such as *glass* and *steel*, have not been mentioned in this book. Try to find out how they are made.

Look at the clothes you are wearing today. What are they made of? Make a list like this: *My pullover is made of wool.*

PART THREE

Looking
at Britain

The publishers are grateful to the following for permission to reproduce photographs:

Aerofilms Ltd 145, 150, 151, 152, 169, 181, 183b, 233b, 234a and b, 236a; Australian News and Information Bureau 195; Barnaby's Picture Library 171, 203, 232a; Brighton Corporation 237a; BAC 175; BMC 173a, b and c, 174; BOAC 176a and b, 227, 228a and b, 229a and b; British Hovercraft Corporation 172b; British Rail, Eastern Region 226b, 233a; British Rail, London Midland Region 223a and b, 224a, b, c and d, 225b, 226; British Resin Products Ltd 177a; British Road Services 220; Maurice Broomfield and Birds-Eye Foods Ltd 208a; John Brown and Co (Clydebank) Ltd 170a and b; BX Plastics Ltd 177b and c; Central Electricity Generating Board 164; Central Office of Information 154a; Cotton Board 182, 183b; Courtaulds Ltd 238b; Cunard Steam Ship Company 231b; Esso Petroleum Co Ltd 172a; Farmer and Stockbreeder 148, 200, 211; Farming World 212; Fox Photos 199; Leonard and Marjorie Cayton 238a; Eric Guy 214a and b; ICI Fibres Ltd 185a and b; International Wool Secretariat 178a, 179; Kent Messenger 204, 205b, 206; The Liverpool Corporation 209, 222b; London Fire Brigade 154b; Luedeke Studio 186a; Milk Marketing Board 201, 202a and b; National Coal Board 161, 162; Geoffrey Pass 219; By courtesy of H.M. Postmaster-General 155a and b, 156, 157a, 167; Director of Public Cleansing, Westminster 158a and b; Radio Times Hulton Picture Library 231a; Rootes Ltd 157b; Roads Campaign Council 221, 222a; Sheffield Newspapers Ltd 159; Leonard Smith 208b; Sport and General Press Agency 237b; *The Times* 196, 199, 232, 235a, 236b; John Topham Ltd 147, 205a, 235b; Vauxhall Motor Company 157b; Josiah Wedgwood and Sons Ltd 187, 188a, b and c, 189a, b, c and d; Michael Wood 216, 218; Yorktown Manufacturing Co Ltd 186

146

CONTENTS OF PART 3

This farmer is ploughing his fields in the autumn

About part 3

This part of the book is about life and work in Britain. It tells you about some of the people of England, Scotland, Wales and Northern Ireland, and shows you the places where they live.

Some of these people live in towns and cities, and work in factories and offices. Some live in the country and work on farms. Some live near the coast, and work as fishermen, or they load and unload ships at the great ports.

As well as telling you about people and places, this part also shows you how maps are made, and it tells you about the weather and the seasons.

When you are reading this part try to think of ways in which Britain is changing. Life today is very different from life fifty years ago. What will it be like in fifty years' time? Will anyone burn coal on open fires? Will airliners need long runways? Will houses be made of bricks and mortar? Will oil be carried about the country in tankers? Will cows graze in fields? Will men work forty hours a week in factories? Will a lorry on a motorway need a driver?

1 Map making

Making a plan of a chair

Peter's chair has a seat which is 300 millimetres (mm) square: each side of the seat is 300 mm long. The seat is too large to be drawn full-size on a small sheet of paper, but this drawing shows what it looks like when seen from above. In the drawing the sides of the seat are 30 mm long. We call this a scale of 1 mm = 10 mm. We can also write this as 1:10.

Using the same scale, the desk has been drawn 60 mm wide and 45 mm from front to back. What size is the desk top in fact?

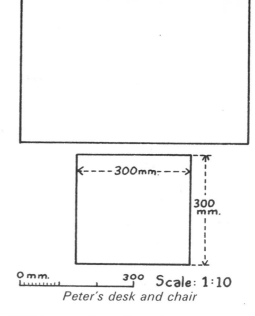

Peter's desk and chair

Making a plan of a classroom

When Peter wants to draw a scale plan of his classroom he finds that the same scale (1:10) is much too large. Instead he chooses a scale of 1:100. This means that 1 mm on the plan represents 100 mm in the classroom.

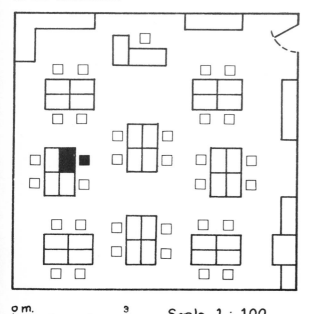

Scale 1 : 100

Peter's classroom, with his desk and chair filled in in black

Making a plan of a school

To draw a plan of his school Peter chooses a scale of 1:500, so that 1 mm = 500 mm. We can also say that 2 mm = 1 metre.

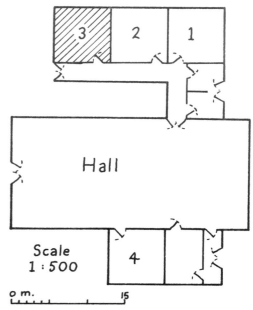

Scale 1 : 500

Peter's school, with his classroom shaded

Map making

Making a map of a small area

Peter lives near the school. When he drew this map of his route to school, he chose a scale of 1:1500 (1 mm = 1·5 m).

Maps of larger areas

When Peter and the rest of his class make a study of their town, they use a map with a scale of 1:25 000. This shows individual buildings, footpaths, woods, parks and many other features in great detail.

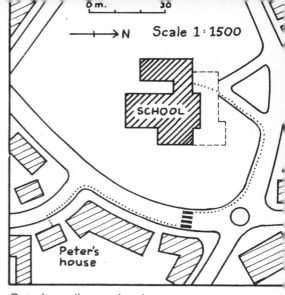

Peter's walk to school.

Car drivers often use maps with a scale of 1:250 000 or, if they are planning a long journey, they study their route on a map with a scale of 1:625 000.

Some maps which are still in use have scales with rather odd numbers in them (1:63 360 is a very common one). This is because they were originally based on a scale worked out in inches to the mile. These maps have metric scales on them and can be used just like any other map.

A seaside town

Here is part of a seaside town, Margate. The Bathing Pool was made by enclosing part of the beach. The large building on the beach is the Marine Pavilion, where teas and ices are served. The dark road behind the beach is the Marine Terrace.

The bathing pool and beach at Margate, in Kent

150

This map shows the Bathing Pool and Marine Pavilion. It also shows some of the houses and other buildings on the promenade.

The map is drawn to a scale of 1 : 2500 (1 mm = 2·5 m). Notice that on a map of this scale, the shape of every building is shown.

Maps are usually drawn with north at the top. The photograph on page 150 was taken by a camera looking north-east.

How many things can you find on this map which tell you that it shows part of a seaside town?

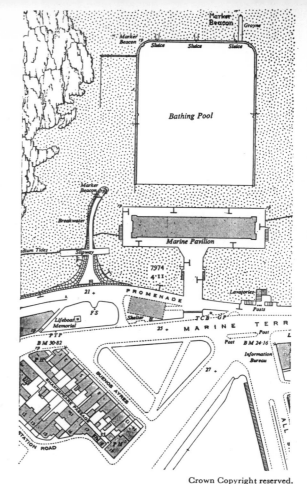

 This little map shows the same part of Margate as the big one, but this map has a scale of 1 : 10 000 (1 mm = 10 m).

Crown Copyright reserved.

Part of Margate (scale 1 : 2500 — 1 mm = 2·5 m)

This is the harbour at Margate as it appears in a photograph and as it is shown on a map with the same scale as the one above (1 : 10 000).

The harbour

151

Margate, looking over the railway station to the bathing pool, harbour and pier

This photograph of Margate was taken from an aeroplane flying towards the north-east. You can see the Bathing Pool, and behind it the Harbour and Pier. Half-way along the Pier is the Lifeboat House.

On the left of the picture are several rows of houses. Behind them, near the sea, is a tall row of hotels and boarding houses.

In the front of the picture is the railway station.

At the top of the opposite page is a map of Margate (scale 1:10 000). The dashed lines (— — —) show the part of the map which is shown in the photograph.

Find the Bathing Pool, the Harbour, and the Lifeboat House.

Find the railway station, and the names of the roads near the station.

What would you see on a walk from the station to the Harbour?

152

A map of Margate (scale 1:10 000)

Margate is in the county of Kent. A map which shows the whole of Kent must be on a small scale. This map of Kent has a scale of 1:2 000 000 (1 mm = 2 km). Margate is shown as a small dot.

A map which shows the whole of Britain must be on a very small scale. This map of Britain has a scale of 1:20 000 000 (1 mm = 20 km).

A map of Kent (scale 1:2 000 000— 10 mm = 20 km)

A map of Britain (scale 1:20 000 000 —10 mm = 200 km)

153

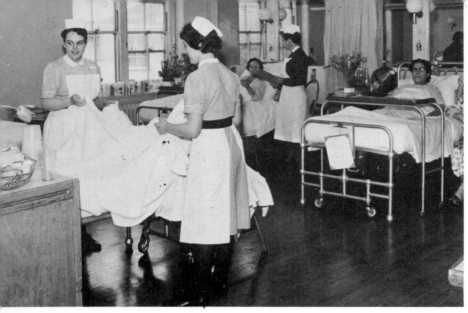

A ward in a hospital

2 A town and its services

Besides the many workers who *make* things—things like shoes, tables, motorcars and peppermints—there are many who *do* things instead—firemen, office workers, policemen and rat-catchers. A surprising number of people provide services, instead of making things.

A civilised country must have many of these services, such as the police force, public transport, libraries and gasworks, if life is to go on safely and comfortably for its citizens. Every day postmen deliver letters, water flows from the kitchen tap, and dustmen empty the dustbins.

Fighting a fire

The post

Posting a letter

This letter is being posted at a post-box in Devon. Notice the postcode (YO1 6EB). This means that the letter can be sorted by a machine. Once the letter goes into the machine, it will be automatically sorted until it reaches the postman at York.

At the sorting office

Postmen in vans collect the letters from the letter-boxes and take them to the local sorting office.

The letters are put into a machine called a separator. This tosses out the heavy packets, which are dealt with by hand. The letters will go through the automatic machines. Another machine turns all the letters round the same way.

The stamps are "cancelled" by printing a postmark over each stamp. The postmark usually gives the name of the sorting office, the date and the time.

Then a man reads the postcode and types the code in "machine language". This is a row of dots, which our eyes can hardly see, but the machine reads easily. The postman

Typing the postcode in "machine language"

Taking the sorted letters from the machine

in the picture is taking the letters from a pigeon-hole marked "Isle of Man". The letters are tied into bundles, labelled, and dropped into mailbags.

Delivering the mail

Post Office vans take the bags of mail from the sorting office to the railway station, where they are put on to fast trains. In London so many letters and parcels are posted and received that the Post Office has its own narrow-gauge underground railway between the large sorting offices and the main line railway stations.

Some trains have a Travelling Post Office, which is a sorting office in a railway coach. Mailbags are picked up by these trains while travelling at full speed. A net on the side of the coach catches the leather mail bag which is hanging from a metal arm alongside the track. Bags are dropped in a similar way.

The travelling Post Office picks up a mailbag in its net

When the mailbags are unloaded at the railway stations they are taken by van to the local sorting offices. There the letters are sorted into the roads and house numbers, so that the postmen waste no time in doubling to and fro when they are delivering the letters.

Many letters are sent to other countries. People who want their letters to travel quickly send them by air. Special light-weight air letter forms are sold at Post Offices.

At the Post Office

You know that you can buy stamps at a Post Office counter. But do you know how many other things can be done there? Look at the list at the top of the next page.

The counter of a large Post Office

156

Here are some things which people can do at a Post Office:

Buy Premium Savings Bonds
    National Savings Stamps
    a Dog Licence
    a TV Licence
    Postal Orders, Money Orders
Renew a Car Licence
Bank money in the Savings Bank
Draw Retirement Pensions
    Family Allowances
Pay a telephone bill
Send a telegram
Post a parcel
Operate a Giro account

From this list alone you can see that the Post Office provides a very useful *service* for the public.

Delivering a telegram

Dustmen at work

For hundreds of years people used to throw their rubbish into the street. Sometimes the Mayor sent round carts to collect the rubbish and tip it outside the town walls. But in the end the streets became so dirty and smelly that the job had to be done properly and regularly.

Nowadays, dustmen call about once a week to empty our dustbins into a dustcart. When the dustcart is full, the rubbish is taken to the disposal depot to be sorted. The rubbish is tipped out on to a moving belt where men pick out the things which can be salvaged — glass, rags and metals. Sometimes magnets are used to remove iron and steel. Salvage is so valuable that in some towns the men are paid extra for collecting it.

Emptying dustbins automatically

Some rubbish is tipped on low-lying land. Barges, lorries or railway trucks carry the rubbish to the tips

Getting rid of the rubbish

Most of the rubbish is burned in incinerators, but in some towns the rubbish is tipped on to waste land. Rubbish cannot be tipped just anywhere. Sometimes it is tipped on low-lying land to raise it, or on to uneven ground to level it. Rubbish which is tipped must be covered with layers of soil, so that there is no smell and so that paper does not blow away. When the tip has been left for several years to settle down, houses may be built there. Sometimes the rubbish is loaded into barges and taken out to sea, where it is dumped. In America and Canada and occasionally in Britain, all rubbish (even tins) is put into a "mincer" which grinds it up so small that it can be washed away down the drain.

Water

You know that the water which comes from the tap once fell as rain. But do you know why we can still get water from the tap even when no rain has fallen for several weeks? Why is tapwater so clean, when water in ponds and rivers is usually very dirty?

A reservoir on Dartmoor, Devon

158

Rain falling on the hillsides forms little streams which may flow into the valleys between the hills to form a lake. Often men make a lake by building a huge concrete wall, called a *dam*, to hold back the water in a valley.

These storage lakes are called *reservoirs*. The water which flows into the reservoir is often muddy, but in the reservoir any mud in the water sinks to the bottom. You can see how this happens if you leave a jar of muddy water standing for a time.

Ladybower Reservoir, Derbyshire, during a drought

There is water in the reservoir even after a *drought*, a long period without rain.

Making the water pure
The clear water in the reservoir is pumped through large pipes to the town. Then it is passed through fine sand, to *filter* it, that is, to take out all the dirt. Often water contains bacteria which are killed by adding to the water a chemical called chlorine. If the water is stored again it is kept in covered reservoirs to keep it pure. The water is constantly tested to make sure that it stays pure.

The pure fresh water is pumped from the waterworks along the "mains" which run underneath our streets. Smaller pipes lead from the main supply to each house.

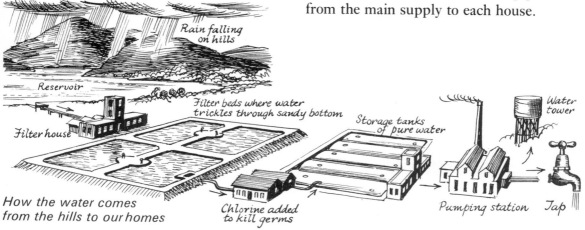

Rain falling on hills

Reservoir

Filter house

Filter beds where water trickles through sandy bottom

Chlorine added to kill germs

Storage tanks of pure water

Pumping station

Water tower

Tap

How the water comes from the hills to our homes

A town and its services

Water from underground

If rainwater falls on chalk soil it soaks through the chalk until it reaches a layer of clay or rock and cannot sink any further. The chalk holds water in much the same way as a sponge holds water. If a well is dug in the chalk it fills with water. The water can be drawn up in a bucket, or pumped up by an electric pump.

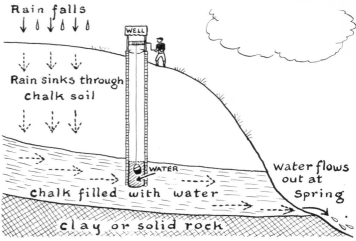

Water sinks through chalk, but cannot pass through clay

Water from under the ground is often so pure that we can drink it just as it comes out of the ground. But usually the water is "hard" that is, it has chalk or limestone in it and it is difficult to get a good lather with soap when we wash our hands. Water from lakes is usually "soft".

130 to 170 litres of water are needed every time an automatic washing machine like this is used

Water for the big cities

London gets most of its water from the valley of the River Thames. The rest it gets from the valley of the River Lea and from wells. The river water is pumped into reservoirs on the outskirts of London before being filtered. The water pumped from the wells in the chalk is so pure it can be pumped straight into the main supply but it is chlorinated as a precaution.

Some towns, and many factories, take their water from rivers, or from lakes in the hills a long way away, where the water is pure. Manchester gets its water from Hawes Water and Lake Thirlmere in the Lake District. The River Elan in central Wales has been dammed to make the Caban Coch reservoir, to provide water for Birmingham 120 kilometres away. Liverpool, too, gets its water from an artificial lake in North Wales. Most of the water for Glasgow comes from Loch Katrine, 50 kilometres from the city.

3 Power for our factories and homes

What makes the wheels of a bicycle go round? When you ride a bicycle, your legs press the pedals which drive the wheels. Your legs are the *power* which makes the bicycle work.

A horse provides the power which is needed to pull a cart. Water power can be used to drive a mill wheel, and even to make electricity. Power is needed to move things, or to make machines work. Railway trains and cars must have power to drive them, and aeroplanes must have power to keep them in the air. Power is needed for heating, for lighting and for cooking.

A great deal of power is needed to supply all the factories, homes, vehicles and shops.

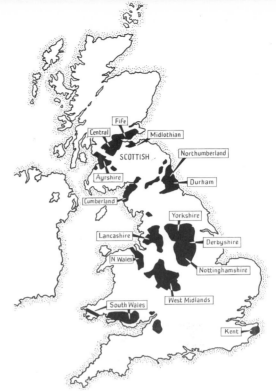

The coalfields of Britain

The excavator moves huge quantities of earth at each "bite"

Coal used to be our main source of power and we still use a great deal. Much of the coal lies far underground, and deep mines have to be dug to reach it. But in places where the coal lies near the surface it can be dug out without going underground. This is called opencast mining. Huge mechanical excavators scoop off the top soil to expose the coal.

When the covering soil has been removed, another excavator digs out the coal. Afterwards the soil is replaced and, in time, the land can be farmed again.

Coal from pits and opencast workings still goes to factories and homes in many parts of Britain. But nowadays the main use of coal is at power stations, to make electricity.

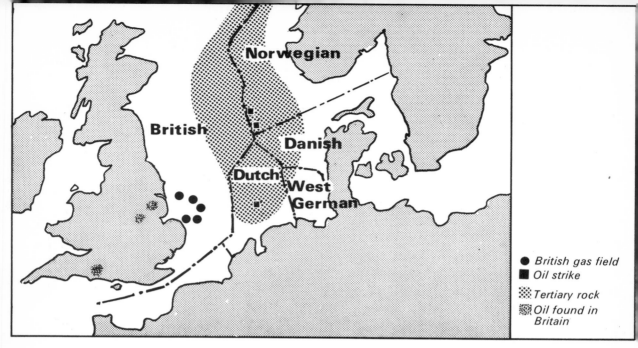

British gas field
Oil strike
Tertiary rock
Oil found in Britain

This map shows the area in which the search is being made for natural gas and oil. Britain is already piping gas from under the North Sea, and hopes that, like the Dutch and Norwegians, she will find oil in her section of the North Sea. Oil is already found in some parts of Britain, and a further search is being made in the Irish Sea.

Oil and gas

Oil is a very important source of power. It is found by drilling through the surface of the earth, either on land or under the sea. The crude oil is treated in refineries to make petrol for cars, kerosene for lighting and heating, diesel oil for heavy engines, fuel for jet engines, fuel oil for heating and power, white spirit for paint and many other products, chemicals, plastics, and gases such as butane, propane and methane.

Nearly all the oil used in Britain is brought by tankers from the Middle East, Libya, Venezuela and Nigeria, and soon we may be getting it from Alaska too where new discoveries have been made. The oil is delivered to the refineries round our coasts or sent inland by road, rail and pipeline for distribution.

Oil refineries in Britain

One of the drilling rigs used in the North Sea

Methane or natural gas is also found under the earth. Recently important discoveries of natural gas have been made in the North Sea and soon all the gas used in Britain will be methane, either from the wells in the North Sea or manufactured from petroleum. Before natural gas can be used to cook the Sunday dinner, the gas stove (and, of course, all the other appliances in the house) must be specially adapted to burn this type of gas. Each area in turn is converted to natural gas and this means that thousands of gas stoves have to be altered on one day.

The natural gas is distributed to the areas which use it by pipeline and new pipelines are being planned to carry natural gas from the east coast up to Scotland and across to Wales and the West Country.

The oil companies are still prospecting for more natural gas in the North Sea and also in the Irish Sea, and they hope that they will find oil as well.

How electricity is made

Electricity is a very convenient form of power. We use it for heating and lighting our homes, to drive machinery in factories and to run trains.

How is electricity made? A dynamo on a bicycle makes electricity. As the bicycle wheel turns, the dynamo spins, so making electricity to light the lamp.

In the same way, on a larger scale, huge turbines at power stations drive generators which make electricity. (A turbine is shown on page 165.) At the power station the turbines are turned by steam power. Coal, oil, or nuclear fuel is used to heat the water to make the steam.

The steam which has driven the turbine has to be cooled. Much water is used for cooling the steam, so power stations are nearly always built near rivers or near the sea, or they have high cooling towers. In London, hot water from Battersea Power Station is supplied to nearby flats.

From the power station the electricity is carried by power cables. Steel towers called *pylons* carry the cables across the hills and valleys. Because electricity can so easily be sent along wires, power stations do not need to be built near to the places where the electricity will be used.

Coal has been mined for so many years in Britain that many of the best seams are already worked out. A time may come when there is no coal left, but we are using less and less each year. We cannot be sure, either, that we shall always be able to import enough oil from other countries. Fortunately there are other ways of making electricity, as well as by using coal or oil.

Hydro-electric power

Where there are hills or mountains, and plenty of rain, electricity can be made by using water power. Electricity made in this

An atomic power station inside the Snowdonia National Park in North Wales. It is the first to use a lake to obtain cooling water

rain

loch

dam

pipe line

power station

dam and
power station

loch

*Rain which falls in the
hills flows into the lochs.
From the loch the water
rushes down pipelines to
the power station.*

way is called *hydro-electric power* (hydro means water). As rain falls on the hills it forms streams and rivers. Engineers build a dam across a valley, so making a lake (called a loch in Scotland). From the lake, the water rushes down a pipeline to a generating station. The force of the water turns a turbine which drives a generator, so making electricity.

There are hydro-electric power stations at Loch Sloy, Pitlochry and other places in Scotland, and in the Snowdon region in North Wales.

Atomic power

Scientists have discovered a wonderful new source of power in the atom. When an atom is split, great heat is given off. This heat can be used to make steam to drive a turbine, so making electricity by *atomic*, or nuclear, power. There are atomic power stations at Sizewell, Calder, Berkeley, Dungeness and many other places.

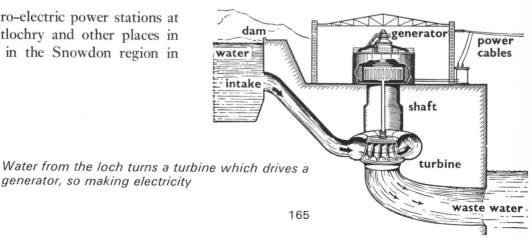

dam

generator

power
cables

water

intake

shaft

turbine

waste water

*Water from the loch turns a turbine which drives a
generator, so making electricity*

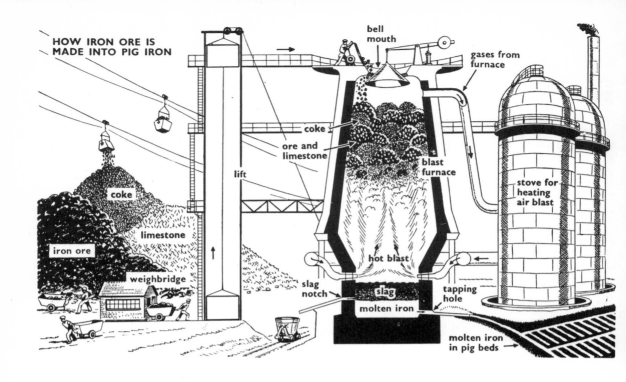

HOW IRON ORE IS MADE INTO PIG IRON

bell mouth

gases from furnace

coke

ore and limestone

lift

blast furnace

stove for heating air blast

coke

limestone

iron ore

weighbridge

hot blast

slag notch

slag

molten iron

tapping hole

molten iron in pig beds

4 Iron and steel

Iron ore and where it is found

Iron ore is a rock which usually lies just under the surface of the ground. When the covering layer of soil has been removed, the iron ore is dug out, and the iron is removed from the ore. Steel is made from the ore; ships, bicycles, pins and cars, are a few of the things which are made of steel.

In Britain, iron ore is found in Northamptonshire, Lincolnshire, Yorkshire and Cumberland. But better ore, which contains more iron, comes to Britain by ship from Sweden, North Africa, Spain, Canada, Venezuela and Liberia.

Smelting the iron from the ore

The ore is mixed with coke and limestone tipped into the top of a tall steel tower, 30 m high. Inside the tower there is a raging fire, which is made white hot by a hot blast of air. The hot air makes the coke burn so fiercely that the iron in the ore melts, and becomes a white-hot liquid, molten iron, which runs down to the bottom of the furnace. Melting the iron from the ore is called "smelting". The tall tower is called a blast furnace.

Every eight hours the liquid iron in the furnace is let out from a tapping hole at the bottom. A stream of white-hot iron pours into "pig beds", or into moulds on a moving belt. In the moulds the iron is sprayed with water to cool it, so that it sets into hard blocks of "pig iron".

Pig iron, cast iron and wrought iron

Some of the pig iron is heated again to make it molten. It is poured into specially-shaped moulds to make *cast-iron* pillar-boxes, lamp posts, grates and other things. Although cast iron is hard, it is brittle, for it breaks easily if it is hammered.

Some pig iron is heated again until it is soft enough to be hammered into shape, to make *wrought iron*. Nails, gates and farm implements are often made of wrought iron.

Steel, and how it is made from iron

Steel is used to make rockets, space craft and radio telescopes as well as cars, bicycles and trains. It was used to make the Forth Bridges, and the liner *Queen Elizabeth 2*. Steel can also be used to make tiny, delicate things such as needles and watch springs.

The aerial of the GPO's radio station at Goonhilly, Cornwall, is made of steel

Most iron is made into steel. White-hot iron from the blast furnace is carried to the steel furnace in a huge ladle lined with firebricks. The iron is heated again with limestone and with scrap metal such as railings and old bicycles. A gas called oxygen is blown into the furnace to speed up the change from iron to steel but even so it takes eight hours before the furnace is ready for "tapping".

This huge shovel, worked by the man on the left, carries scrap metal to the furnace

167

Pouring molten metal into ingot moulds

Large ingots, of high-quality steel, are worked under a steam hammer.

Many of the things which are done in a steel works are controlled by machines. Steel making is one of Britain's most highly automated industries.

Britain's iron and steel works
Iron and steel works used to be built near coal mines. Then they were built near iron mines. But now huge quantities of ore are needed, and because this comes by ship many steel works have been built on or near the coast. There are big iron and steel works in Lanarkshire, on Tees-side, in South Yorkshire and in North and South Wales.

Tapping the steel furnace
When the furnace is tapped, a river of white-hot molten steel pours out into a giant ladle, and showers of sparks fly in all directions. From the ladle the molten steel is poured into huge moulds. When the steel cools it can be taken from the moulds as a solid lump, called an *ingot*. The ingot may weigh as much as thirty tonnes.

Working the steel ingot
The steel ingot is then worked into whatever shape is needed: it may be a bar, a flat plate, or a girder. This is done by hammering, rolling or pressing.

Many ingots are flattened in a "rolling mill". There is a deafening noise in the mill as the red-hot ingot goes backwards and forwards between the rollers until it becomes a long flat slab of steel. Later, the slabs can be rolled again into strip steel which is cut up into sheets of steel for making car bodies and tin-plate.

A steam hammer, striking an ingot

Shipyards on the Clyde

5 Building a ship

Because Britain is an island, we have always needed ships. We have built them not only for ourselves but for other countries too. Today, many of the ships sailing into harbours all over the world were built in British shipyards.

Modern ocean-going ships are made almost entirely of steel. They are built on the banks of deep wide rivers, where they can be launched safely, and "fitted out" or completed, without being exposed to the rough sea. Shipyards are always near steelworks, for steel is very heavy and it is costly to carry it far.

Planning the ship

When a new ship is wanted, the men who are going to make it say to themselves:
"What cargo will it carry?"
"How big should it be?"
"Which ports will it use?"
"What sort of engine should it have?"

Then designers set to work, making hundreds of drawings of every part of the ship. When they have planned the ship, an accurate model of the *hull*, called a shell, is made. The model is tested in many ways in a water tank, to make sure the design is good.

Joining steel plates together by using an arc-welding set

The riveter's hammer is driven by compressed air

After many months of preparation, cranes are put into place on the slipway where the ship is to be built, and work begins on the hull. Firms in many parts of the country will make the parts of the ship: the engines, the steel plates, the propellers and all the other fittings.

Building the ship

First the *keel*, or backbone of the ship is laid. It is made of huge steel girders which are joined together by *rivets*. A rivet is a small piece of red-hot steel which is put through holes in the two girders and then hammered. As the rivet cools, it pulls the girders together.

A framework of girders, looking like ribs, is built on to the keel, and covered with steel plates, which are riveted on to it.

Sometimes the plates are *welded* together. This means that the edges of two plates are heated and made so hot that when they are hammered together they become one.

Nowadays many smaller ships are *prefabricated*. This means that parts of the hull are built separately in workshops and then welded together on the slipway.

When the hull of the ship is complete, there comes the most thrilling moment —the *launching* of the ship.

Fitting out

Tugs tow the ship to a *fitting-out basin*, nearer the mouth of the river, where the engines and all the fittings are added. It takes two years of hard work before a big ship is ready to sail the seas.

170

When the ship has been fitted-out, her makers take her out "on trial". At sea she has to pass tests to prove that she will do all that her designers planned.

A refit in dry dock

After a ship has been sailing for some time she must have an overhaul, or re-fit, as it is called. Nearly every port has a repair yard, and many have a *dry dock*. When the ship sails into a dry dock, gates are closed behind her. The water is pumped out of the dock so that the hull of the ship can be repaired and painted before its next voyage.

Britain's shipyards

The two most important shipbuilding areas in Britain are the north-east coast of England and the estuary of the River Clyde.

In north-east England there are shipbuilding yards near the mouths of the rivers Tyne, Wear and Tees.

One of the largest shipyards is Harland and Wolffs', at Belfast. There are many

A ship in dry dock at Liverpool for an overhaul

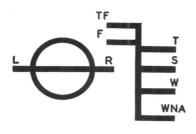

The Plimsoll line, painted on a ship's side, shows how deep in the water the ship can be loaded. The depth varies with the season and the kind of water. (LR=Lloyd's Register; T=tropical; S= summer; W=winter; WNA=winter North Atlantic)

shipyards on the River Clyde in Scotland, at Dumbarton, Glasgow and Greenock.

Cammel Laird have large shipyards at Birkenhead on the River Mersey, where they build cargo ships, liners and tankers. Barrow is famous for building submarines.

Shipyards along the estuaries of the Rivers Forth, Tay, Tyne and Tees, build cross-channel steamers, and "tramps" (ships which carry any cargo and sail on no fixed route). Fishing boats are built at many of the fishing ports, such as Aberdeen and Hull.

Esso Northumbria *is 350 m long, and has a beam of approximately 50 m*

A Mountbatten class Hovercraft

Many ships built in British shipyards are for special tasks: trawlers, drifters, whalers, tugs, dredgers, lightships and cableships. Refrigerated ships carry meat, fruit, butter, eggs and fish.

Oil tankers

The largest ships built in Britain today are oil tankers. A tanker has only one deck, and its engines and living quarters are near the stern. The bridge is amidships, with a gangway called a "catwalk" connecting it to the forecastle and stern.

A tanker like *Esso Northumbria*, which was built in the Wallsend yard and launched in 1969, carries 25,000 tonnes of crude oil. Each of the many tanks in which the oil is stored has its own pipes, so that the ship can be unloaded in a few hours.

The ship's engines

Most of the ships sailing today have a screw, or propeller, to drive them. The propeller is turned by an engine. For many years most ships were driven by coal-fired steam engines, but many of these have been converted to burn oil. Nearly half the ships sailing today, and most newly-built ships, are driven by diesel engines. Nuclear power is used to drive some submarines.

6 Making cars and other things from iron and steel

Steel from the steel works goes to factories in many parts of Britain. A great deal of it goes to factories making motor vehicles. Did you know that over two million vehicles are made in Britain every year? As well as steel, the finished cars and lorries need many other materials, such as rubber, glass, plastics, leather, wood and aluminium.

The parts of a car are often made in different factories—the engine in one factory, the electrical fittings in another, the tyres at another, and so on. Then all the parts are brought together at one factory and *assembled* to make the finished car. Parts from 2,000 firms are needed for an Austin 1800.

The "assembly line"

The basic parts of the car (such as the engine and the body shell) are built up separately and then sent by conveyors to the final assembly line. Here they are bolted together and the complete car takes shape.

Each man on the line does the same job all day, fitting or adjusting one particular part of each car when it reaches him. In this way he becomes very skilled at the job he is doing, and does not waste time in changing from job to job.

In the control room a man supervises the flow of the assembly line

173

A Mini body comes out of the paint spray booth

The finished body is lowered on to the engine and chassis and bolted down

Britain is famous for making small cars, motor cycles and magnificent large cars which are used by ambassadors, officials and business-men all over the world. One in four of the vehicles made in Britain is sold abroad.

Where cars are made
Most car factories are in the Midlands and in the London area. There are factories at Birmingham (making *Austin* cars), at Coventry (*Jaguar*, *Hillman*), Oxford (*Morris*), Crewe (*Rolls Royce*), Luton (*Vauxhall*), and Dagenham (*Ford*).

"Heavy engineering"
Glasgow, Lanarkshire, and Middlesbrough are famous for "heavy engineering". Ships are built there and railway engines, cranes, lifts, bridges and lathes.

"Light engineering"
Light engineering works make small things from iron and steel, and so the works themselves do not need to be large. They make typewriters, watches, electric fires, etc. There are many light engineering works in the Midlands of England, around Birmingham.

In some factories steel is coated, or *plated*, as it is called, with another metal: tin-plated steel is made into tins to store meat, fruit and vegetables. Some steel is plated with zinc, or *galvanised*, to make dustbins, buckets, and corrugated sheeting for roofs. Steel and plastic can be bonded together for use in refrigerators, water heaters and cars.

Tools for the job
Many of the tools used in the factories of Britain are made in Sheffield. Iron ore was once mined near Sheffield and coal has always been plentiful there. But it was the hard local stone, good for making grindstones, which first made Sheffield famous for sharp tools. All over the world people rely on tools marked "Made in Sheffield".

Ploughs, harrows, and all the other tools which a farmer needs, must be strong and lasting, so they are made of iron and steel. Many farm implements are made in small factories in market towns such as Norwich, Shrewsbury, Ipswich and Lincoln.

This car transporter makes daily runs from Birmingham to Manchester via the M6

The huge prototype of Concorde, *designed to fly at supersonic speeds, was the result of many years of research and experiment*

7 Building aircraft

An aeroplane is made in a similar way to a motor car. Many of the parts of the plane are made in factories in different parts of Britain and are later assembled by the "parent" company. But aircraft cannot be produced in the same way as cars. Every part must be carefully inspected if the plane is to be safe and airworthy.

Aircraft factories

An aircraft factory usually has its own airfield. But because heavy modern aircraft need a runway nearly two miles long for taking off and landing, factories can only be built where there is plenty of flat land. There are large aeroplane factories at Hatfield, north of London (where the Hawker Siddeley *Trident* is made), at Bristol (the *Concorde*), at Weybridge in Surrey (the *VC 10* and at Kingston-on-Thames, the headquarters of the Hawker Siddeley Group.

Many kinds of aircraft are built today: airliners, military aircraft and helicopters, as well as planes for such work as surveying, freight carrying, and spraying crops. A modern airliner may cost as much as two million pounds to buy.

Planning and designing the aircraft takes many months, sometimes even years. When production begins, the parent company makes the frame, or skeleton, of the plane, and the sheet metal skin which covers the frame.

Building aircraft

All the other parts needed for the aircraft are specially made by a vast number of firms who make electronic equipment, switches, brakes, tyres, windows, seats and so on. Rolls-Royce and Bristol Siddeley are makers of famous engines.

The first aeroplanes had to be very light, because the engine had so little power. Now that engines are so much more powerful, some planes are as heavy as 140 tonnes, but even so the planes must be kept as light as possible. (Heavy planes need much fuel and long runways.) Only very light metals such as aluminium, magnesium or a new metal called titanium, can be used, or mixtures, called *alloys*, of these metals. Steel is used for those parts of the plane which have to stand great heat.

When the frame of the plane is complete the different parts and "fittings" arrive by lorry at the factory, and gradually the plane is completed.

The controls of a VC10 airliner

The test flight

At last the aircraft is ready for its test flight. At the controls is the Chief Test Pilot of the company. When the plane has been given its Certificate of Airworthiness it is delivered to the firm which has ordered it.

The manufacturers make aircraft not only for our own airlines and for the Royal Air Force and Fleet Air Arm, but also for countries overseas.

Most years an exciting flying display is given at Farnborough, when new aircraft are seen for the first time. Buyers from all over the world visit it to see and to buy the aircraft which are on show.

Airliners even have a small kitchen, or "galley"

176

8 Plastics

How many things in your house are made of plastic? Here's a list of plastic things found in one house:

cups	records	wallet	hosepipe
saucers	telephone	toys	watering can
handbag	table top	macintosh	

"Plastic" is a word like "metal" or "wood": it can mean any one of many different materials. But all plastics are similar in that they are light in weight, they don't rust, and they can be coloured all through.

Plastics are made by mixing chemicals (powders, liquids or granules). Because they are difficult to make they are usually made in machines which never stop: it would not be worth while to mix small quantities of a plastic, so factories work day and night and the workers work in "shifts".

There are three ways of making plastics into shapes that are needed:

moulding—by heat and pressure in a steel mould (to make bowls and plates)

extrusion—by forcing heated material through a die (to make tubes)

vacuum forming—by placing a plastic sheet over a shape, and then sucking out the air under the sheet (to make fancy boxes).

Because plastics have many uses the industry is found in many different parts of Britain—wherever there is labour, and space for buildings.

Here are the names of some plastics. Can you find out what is made from each of them: acrylics, polystyrene, polythene?

Christmas decoration made by vacuum forming

Taking a plastic washing-up bowl from the moulding machine

Weaving deckchair fabric. The yarn is made by extrusion

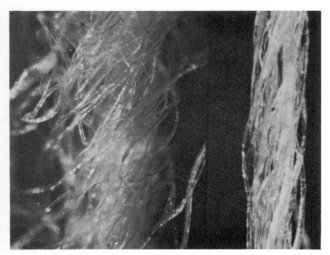

Woollen thread. Worsted thread. Both are seen through a microscope

9 Wool

Why are you warm when you wear woollen clothes? Look carefully at one thread of your woollen pullover or jersey. You will find that it looks like the thread on the left.

Each thread is made up of tiny hairs which were once the hairs, or wool, of a sheep. The air between the hairs keeps in the warmth of your body, and that is why you feel warm when you wear wool.

Thousands of years ago early man used to kill a sheep and skin it. Then he would wear the skin just as it was, with the wool still attached. Coats and sheepskin boots are still made in this way today.

Later, men learnt that the wool could be cut from the sheep and woven into cloth. When the wool had been shorn from the sheep, it was washed, and when it was dry it was combed. The short fibres of the wool were *spun* by the "spinsters" of the household into a long thread called *yarn*. The yarn was then woven into cloth on a machine called a loom.

The people who lived in the Pennine hills kept many sheep, and washed the wool in the soft water of the streams. The wool which they did not use themselves, was sold in the towns, or to travelling merchants.

Shearing the wool from a sheep. The fleece has to be taken off in one piece

178

As the population of Britain grew, and more people wanted woollen clothes, machines were invented to spin and weave the wool more quickly. These machines were too big and too costly for most people to buy, so factories and mills were built. Here, rough wool bought from farmers in the district was spun and woven into cloth. The mills were always built near a stream so that the water of the stream could be used for washing the wool and for driving the machines in the mills.

Later, when steam engines were invented, they were used to drive the machines. Coal was needed to make the steam engines work, so in time mills were built in places where there was wool, coal and water. Many mills were built in the West Riding of Yorkshire to weave the wool from the sheep which grazed on the Pennine hills.

Wool from other countries
As the population continued to grow, more and more wool was needed. There were not enough sheep in England, so wool was obtained from other countries: from Australia, New Zealand, South America and South Africa. Today, wool from these countries comes to London, Liverpool and Hull, and it is sent to the mills by road.

At the mill the wool is sorted according to its length, its strength, and how "crimpy", or wavy, it is.

The wool is still just as it was when it was taken from the sheep: dirty, greasy and tangled. It is boiled with soap, and dried by hot air.

The wool is then combed by spikes on a roller. If the wool is to be made into fine "worsted" cloth, it is combed again until all the hairs lie the same way.

This comb pulls out the long fibres from the short

179

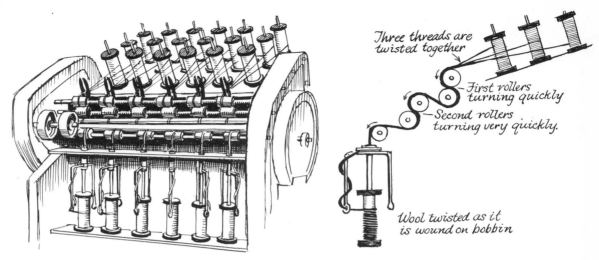

Three threads are twisted together

First rollers turning quickly

Second rollers turning very quickly.

Wool twisted as it is wound on bobbin

The spinning machine pulls out and twists the loose slivers of wool. This makes a thin yarn

Spinning the wool

The combed wool is drawn off as a long loose rope called a *sliver*. The slivers are too thick and loose for weaving, so they are pulled tight and twisted on spinning machines. This makes a *yarn* which can be wound on to a spool.

Some wool is dyed and sold for knitting wool. Wool which is used for making coats, skirts and suits is woven into cloth.

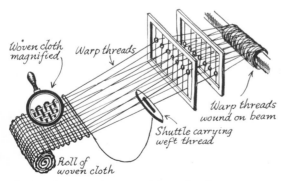

Woven cloth magnified

Warp threads

Warp threads wound on beam

Shuttle carrying weft thread

Roll of woven cloth

A simple weaving machine. As the warp threads are raised and lowered, the weft thread goes backwards and forwards

Weaving the threads into cloth

In simple weaving, one thread goes backwards and forwards, under and over the threads which lie in the opposite direction. But modern looms can weave much more complicated patterns than this.

The cloth which comes from the looms is often fluffy. Cloth for blankets is brushed to make it even more fluffy, but most cloth is beaten to make the threads lie close. Then it is put through a machine which works like a lawn-mower, and cuts off any long hairs. The cloth is damped and left to shrink, and it may be mothproofed before it is pressed and rolled.

Sometimes the woollen yarn is dyed, and different colours of wool are woven to make a patterned cloth. Sometimes the wool is dyed after it is woven, so that the cloth is all one colour.

Halifax, a woollen town in the valley of the River Calder, Yorkshire

The woollen towns of Yorkshire

Most of Britain's woollen towns are in Yorkshire, in the valleys of the rivers Aire and Calder. They are each famous for a different kind of cloth. Carpets are made in Halifax; fine woollen cloth, called *worsted*, is made in Bradford; and a cloth called *shoddy*, made from woollen rags, is woven in Dewsbury and Batley. Both woollen and worsted material are made in Huddersfield. Leeds, the biggest "woollen town", has many factories where cloth is made into suits and coats.

Some other woollen towns

Wool is made into cloth in other parts of Britain, as well as in Yorkshire. In the valley of the River Tweed, in the south-east of Scotland, fine woollen socks, jumpers, and cloth called "tweed" are made.

A good quality cloth, which is used for uniforms and billiard tables, is made at Stroud, in the west of England. Blankets are made at Witney and carpets at Kidderminster and Axminster.

Although nearly all cloth is now made in factories, there are still a few people who spin the wool, and weave the cloth, in their own homes. In the islands of the Hebrides people make the famous "Harris tweed".

British clothes made of wool are among the finest in the world and many of them are sold to other countries. Canada and the U.S.A. buy some of our best woollen goods.

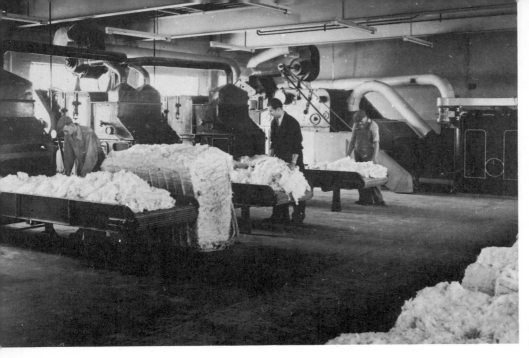

Breaking open the bales of cotton

10 Cotton

The cotton you use every day—as towels or clothes or sewing thread—started its life as a "cotton boll". The "boll" is the seed pod of the cotton plant. When we say "cotton", we may mean the cotton boll, the cotton thread which is used for sewing, or cotton cloth. In this chapter, "cotton" means *cotton cloth*.

In the Middle Ages, the people of Britain usually wore clothes made of wool or linen. Later, cotton was brought from India, but the cotton threads, which were spun by hand, were not very strong. Good cotton was not made until machines were invented to spin strong threads, and to weave the threads into cloth. Then raw cotton was brought from America and landed at Liverpool in Lancashire.

As time went on, Lancashire proved to be the ideal place for the manufacture of cotton. There was water power, and later coal, for driving the machines. The air was damp so that the threads did not break during spinning. The water in the streams was soft for dyeing. Also, the people were skilled at making woollen cloth, and soon learned how to make cotton.

Most of Britain's raw cotton still comes from the United States of America. The Sudan, Brazil, Uganda and West Africa also send raw cotton to this country.

Most of the raw cotton is landed at Liverpool or taken up the Manchester Ship Canal to be unloaded at Manchester. Then it is sent to the mills of Oldham, Bolton, Rochdale and other towns in South Lancashire, where it is spun. Some spinning is done just south of Glasgow.

Spinning and weaving

In the mills different raw cottons are mixed, and then beaten until all the pieces of stalk, leaf or seed have been shaken out. When the raw cotton has been cleaned it makes a fleecy sheet like a huge roll of cotton wool. This goes through a "carding" machine which combs out the fibres. The fine web of fibres is drawn up to make a loose rope called a *sliver*. The sliver is pulled tight and twisted, by a spinning machine, to make a strong thread.

Usually the raw cotton is spun at one mill and the thread is woven into cloth at another. Burnley, Blackburn and Preston are towns in the north of Lancashire which are famous for their weaving.

The carding machine combs out the fibres to make a loose rope called a sliver

If you visit a weaving mill there is so much noise from the rows and rows of looms that you cannot make yourself heard. But the women who work the looms become used to the noise, and learn to lip-read so that they can talk to each other.

A woman weaver can look after sixteen to twenty automatic looms. If a thread breaks, the loom stops itself. Then the weaver joins the thread again and restarts the machine.

Bolton in Lancashire

"Finishing" the cotton

Cotton which is spun in one town and woven in another, may be sent to a third town to be bleached, dyed and finished. Many of the soaps and chemicals needed for these processes are made in Cheshire. The cloth is thoroughly washed and bleached to turn it from grey to sparkling white. Then most of the material is printed either by screen printing (which is like stencilling) or from a rubber roller which picks up dye from an engraved copper cylinder.

Some of the cotton thread is dyed before it is woven, and different colours of thread are woven to make a patterned cloth. Other cotton is dyed after weaving, to make cloth of one colour.

Printing cotton material on a screen printing machine. Where the wax has been removed from the screen, ink is squeezed through on to the material

At one time a great deal of cotton cloth was sent to India and Pakistan. Now the people of these countries have their own cotton mills which make cloth from the cotton grown locally. The workers in the mills are paid very little, so the cloth they make is cheap—much cheaper than the better quality cotton made in Lancashire.

Cotton is popular as a fibre because it is comfortable to wear, it is clean, it is strong, and it is easy to sew. In addition cotton can be permanently pleated, glazed, made shrink resistant and flame resistant. It is often blended with man-made fibres.

Designing a pattern to be printed on cotton

Linen

Linen, a strong cloth used for tablecloths and handkerchiefs, is made from a plant called flax. When the flax is fully grown it is between 1 and 1·30 m tall. The stems are pulled up by hand, tied in bundles, and soaked in water for ten days until the woody parts rot away: this is called *retting* the flax. The linen fibres which remain are cleaned, spun, and woven into cloth.

Linen is made in factories in Northern Ireland. At one time all the flax was grown locally, but now most of it comes from Belgium and Russia.

A climber using a nylon rope *The tyres contain nylon cord, for extra strength*

11 Man-made fibres

Today, many fibres which are used for cloth are made in the same way as a spider makes the thread for its web. A sticky liquid is forced through very fine holes. As the liquid dries it makes a thread or filament. These filaments are twisted together to make a yarn which is later woven into cloth.

Fibres made in this way are often called man-made, but really we mean that they are "made by man from animal or vegetable sources". For example, *rayon* comes from a liquid which is made from wood pulp or cotton waste, both of which once grew in the earth. *Nylon* comes from coal which was once a forest. *Terylene* is made from oil which comes from the remains of tiny creatures buried millions of years ago.

Factories in Coventry, Derby, Lancashire and North Wales make rayon. Nylon is made at Pontypool and Doncaster. Terylene is made in a factory near Middlesbrough.

The finished product
Very often man-made fibres are used alone. Sometimes they are mixed with cotton so that you can have a garment that is comfortable to wear, thanks to the cotton in it, but yet keeps its pleats or creases because of the man-made fibre in it. Both cotton and man-made fibres are very often spun and woven by the same mills. Many mills in Lancashire do not now spin and weave cotton only; they spin and weave whatever fibres happen to be most in demand. There are now some very large firms which own mills in many parts of the country. These firms carry out all the processes of spinning, weaving, finishing and selling the finished cloth.

12 Making clothes

A tailor-made coat

If you want a new coat, you may know someone who can make it for you. He will measure you carefully, cut out the material, and sew the pieces together to make your new coat.

"Ready-made" coats

But more often you will buy a "ready-made" coat from a shop. Hundreds of coats of the same style and material are made in factories. All the cutting-out and sewing is done by machine. In each part of the factory different parts of the coats are made —hundreds of sleeves in one workshop, hundreds of collars in another.

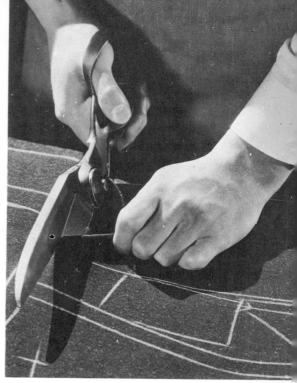

Cutting out a suit by hand

Using a machine to cut 40 blouses

Ready-made coats, manufactured in a factory, are cheaper than coats made by a tailor. They can be produced cheaply because hundreds are made at the same time, and because most of the work is done by machine, and not by hand.

Most clothes are made at factories in large cities such as London, Glasgow and Birmingham. In these places there are many people wanting to buy the clothes which are "mass-produced".

Some clothing factories are near the mills where the cloth is made. The factories of Leeds make suits and coats from woollen cloth woven in other towns in Yorkshire. The factories of Manchester make cotton dresses and shirts from cotton woven in the mills of Lancashire.

13 Making pottery

Pots can be made in many different ways: by rolling the clay into long strips and coiling the clay round and round; by pressing out the clay with one's thumbs; by cutting out slabs of clay and pressing them together; or by pressing the clay into a mould. Pots can also be made by putting a lump of clay on to a revolving wheel. As the wheel spins, the potter skilfully hollows out the centre of the clay with his fingers and pulls up the clay to make the sides of the pot. The pots are baked in an oven called a *kiln*, until the clay is hard.

Many years ago people in the Midlands started to make coarse pottery, called *earthenware*, from the clay which they found near their homes.

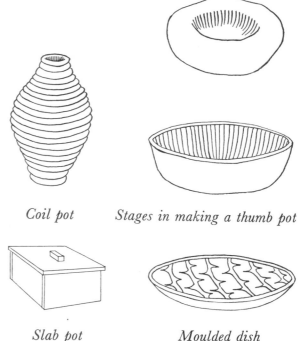

Coil pot *Stages in making a thumb pot*

Slab pot *Moulded dish*

"The Potteries"

As more and more pottery was wanted, the clusters of homes and kilns became villages, and the villages became towns. Stoke-on-Trent is the biggest pottery town in a district of Staffordshire known as "the Five Towns" or "the Potteries". Most of the pottery made in Britain today comes from this area.

Some coarse local clay is still used in the Potteries to make drainpipes and tiles, but most of the clay comes from other places: *ball clay* from Devon or Dorset, china clay or china stone from Cornwall, flint from the sea shore, and bones from Argentina.

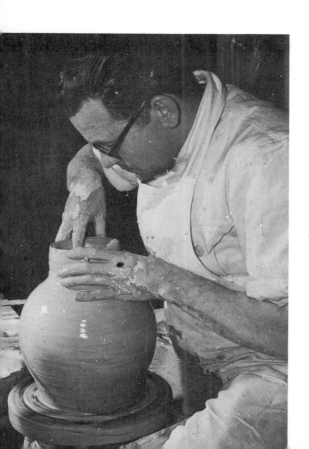

Making a vase on a wheel

Taking a jug from a mould

Mixing the clay

At the pottery these materials are ground into fine powder and mixed with water, to make a sticky clay. The clay is shaped on a wheel to make lovely vases, jugs and tea-pots. But plates, cups and saucers, which are needed by the million, are made in moulds. To make a plate, a ball of clay is flattened on a revolving wheel. The flat clay is thrown into a mould which shapes the front of the plate, and then spun on a "jigger", where the back is shaped by a piece of metal.

Firing the pottery in a kiln

When the clay has been shaped on a wheel, or in a mould, it is baked, or *fired* as it is called, inside a tunnel kiln. The pottery is placed on trucks which move slowly through the tunnel, getting gradually warmer until they reach maximum heat. Then they begin to cool until at the other end of the tunnel they are cool enough to handle.

Some potteries still use bottle-shaped kilns, heated by coal fires. Modern kilns are heated by gas or electricity and do not give off smoke.

Sometimes the clay is sent to the Potteries by rail. But usually the clay, the stones, the flints and the bones go by ship to the River Mersey. Then they go by barge along the canals to the Potteries of Staffordshire.

Shaping the back of a plate

The truck will move slowly through the kiln

Painting a design on a plate

Putting on a paper transfer

Packing the pottery in a barrel

Decorating the pottery

When the pottery has been fired, it is still whitish in colour, so it must be coloured. The finest china is painted by hand, but most china is given a pattern from a "transfer"; the pattern is printed on the paper but can be "transferred" by rubbing on to the china.

When the pots have been decorated, they are dipped in a thin mixture known as glaze. Then they are "fired" again to harden the glaze.

The finished china is carefully packed between straw in barrels and boxes, and taken away from the factory by lorry.

Some famous British firms

Nearly half the china made in Britain is sent abroad. Many people in the U.S.A., in Canada and in Australia buy bone china or earthenware made by such famous British firms as Wedgwood, Spode, Royal Worcester, Royal Doulton, Minton and Royal Crown Derby. Next time you use a plate or a cup, look at the bottom of it. It may say what kind of pottery or china it is, and where it was made.

Engraving a printing cylinder. The cylinder will be "inked" and rolled to reproduce the pattern

189

14 Britain's industry

Britain is an industrial nation: eleven people work in mining, manufacturing and building for every one who works on a farm. There are eight main areas where this industrial work is done.

London
The Midlands
Yorkshire } in England
South-East Lancashire
Tyne-side and Tees-side
South Wales
Central Scotland
and Belfast in Northern Ireland.

Why have most of Britain's factories been built in these areas? There are a number of reasons. Here are some of them.

Coal
Until recently, most factories needed coal to drive their machinery and so the great industrial areas grew up on, or near, coal-fields. Today other sources of power are taking the place of coal, so it is not so necessary for factories to be near to a coal-field.

Raw materials
Many factories are built in the places where their raw materials are found, or where they can easily obtain their raw materials. Thus, they must be close to a main road and railway, and if they depend on materials from overseas, near to a port.

Labour
In some parts of the country there are towns which have only one main industry. For instance in the Rhondda Valley in Wales it is coal mining, in Jarrow it is ship-building, in Oldham it is cotton spinning. If the factories in these towns cannot sell the things they make, many people are out of work.

Nowadays, to help to find jobs for everyone, manufacturers are encouraged by the government to build new factories in areas where there are many men out of work.

Some cities are too big
"New towns" have recently been built, to attract some of the people living in the big cities. These towns have a wide variety of industries.

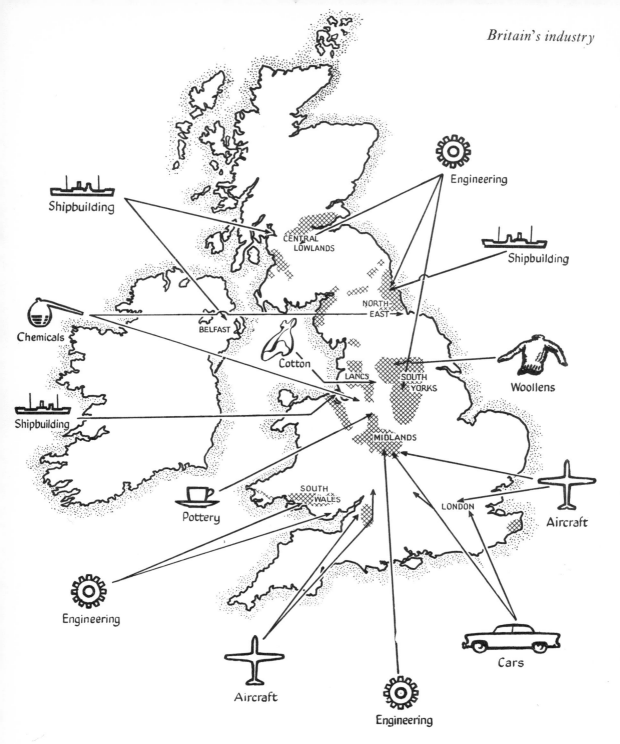

Britain's factories. There are eight main areas where factories have been built. All of them, except London and Belfast, are on or near coalfields. (The coalfields are shaded.)

15 Buying and selling

Britain's imports

Food

Britain's farmlands and farming methods are among the best in the world. But our farms can produce enough food for only half of the people.

Our farms produce almost all the eggs, milk, potatoes and vegetables that we need, but only about half the cheese, bacon, meat, butter and other fats. The rest must be *imported*, or brought from other countries. About four-fifths of the sugar and wheat we need come from other countries.

Some of the foods we like, such as tea, bananas, rice and coffee, grow only in hot countries. These foods, too, have to be imported.

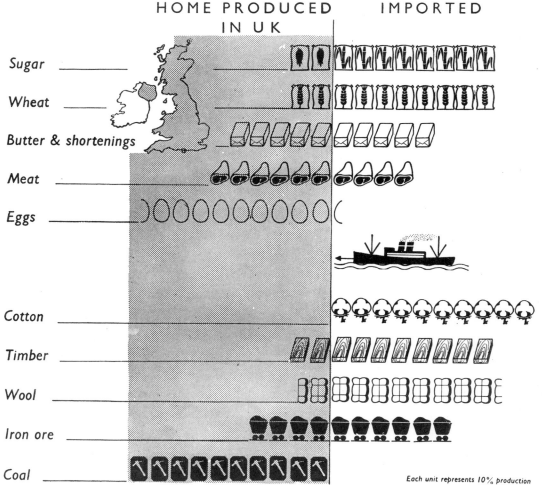

Each unit represents 10% production

Some of the raw materials used in Britain. The diagram shows how much of each must be imported

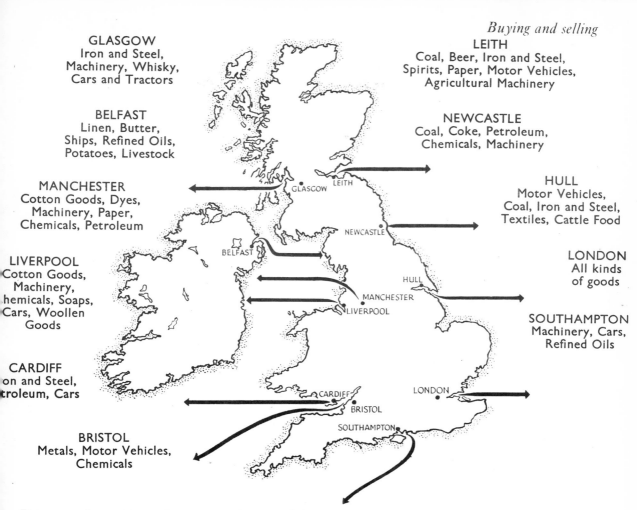

GLASGOW
Iron and Steel,
Machinery, Whisky,
Cars and Tractors

LEITH
Coal, Beer, Iron and Steel,
Spirits, Paper, Motor Vehicles,
Agricultural Machinery

BELFAST
Linen, Butter,
Ships, Refined Oils,
Potatoes, Livestock

NEWCASTLE
Coal, Coke, Petroleum,
Chemicals, Machinery

MANCHESTER
Cotton Goods, Dyes,
Machinery, Paper,
Chemicals, Petroleum

HULL
Motor Vehicles,
Coal, Iron and Steel,
Textiles, Cattle Food

LIVERPOOL
Cotton Goods,
Machinery,
Chemicals, Soaps,
Cars, Woollen
Goods

LONDON
All kinds
of goods

SOUTHAMPTON
Machinery, Cars,
Refined Oils

CARDIFF
Iron and Steel,
Petroleum, Cars

BRISTOL
Metals, Motor Vehicles,
Chemicals

This map shows the main ports, and some of the things shipped from them

Raw materials

All the coal needed in our factories is mined in Britain. But many of the raw materials used in the factories must be imported: over half the iron ore, and most of the wool and timber, come from other countries. (See the diagram on the opposite page.)

Cotton and rubber cannot be grown in Britain and must be imported. Other countries supply all our petroleum, and nearly all our tin, copper and bauxite.

Britain's exports

Schoolchildren are always "swopping" things— stamps for conkers, comics for marbles. In the same way countries "swop" the things they make, or the things they grow, for the things they want.

The diagram on this page shows the most important things which Britain sends abroad ("exports") in exchange for the food and raw materials she imports. It also shows the main ports, and the things which are shipped from them.

16 Day, night and the seasons

The sun gives light and warmth to the earth

The sun is a gigantic mass of flaming gases, more than a million times the size of the earth. The heat of the sun would scorch up anything near to it. But the earth is 93 million miles from the sun. It is far enough away not to be scorched to a cinder, yet near enough to receive the light and warmth which make life possible. Without the sun the earth would be so cold that no living thing could survive, and there would be complete darkness.

The earth spins round once in every twenty-four hours. The sun can only light the side of the earth which faces it, so the rest of the earth, away from the sun, is in darkness. During twenty-four hours each part of the earth has one day and one night.

When there is daylight in Britain there is darkness in Alaska; when there is darkness in Britain there is daylight in Alaska.

When there is daylight in Britain (B) there is darkness in Alaska (A)

The earth goes round the sun

As well as spinning on its own axis, the earth makes a long journey round the sun. The time taken by the earth to travel once round the sun is one *year*.

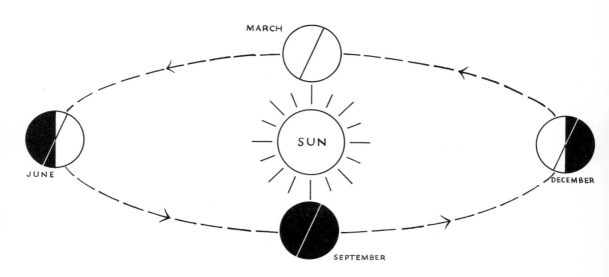

The earth travels round the sun. The time taken on this journey is one year

A torch gives more light when shining directly on a book than when shining at a low angle.
In the same way the sun is hottest when it is overhead

The earth is tilted as it spins

If a torch shines straight on to a book in a darkened room the page is brightly lit. But if the torch shines at a low angle the page is dimly lit, because the same amount of light is spread over more of the page. In the same way the sun's rays may strike the earth directly from overhead or at a low angle.

The sun gives most heat when it is directly overhead, so countries at the equator are hotter than countries nearer the Poles. But as the earth travels round the sun it is not spinning upright, but is tilted. *From March to September the northern hemisphere is tilted towards the sun.* Then the sun is overhead in countries a little to the north of the equator. Britain has its summer, and Australia has winter, for the southern hemisphere is tilted away from the sun.

From September to March the southern hemisphere is tilted towards the sun. Then it is winter in Britain and summer in Australia. Christmas Day in Britain is usually cold. But some Australians have their Christmas dinner on the beach, sitting in the sunshine.

From September to March the southern hemisphere is tilted towards the sun

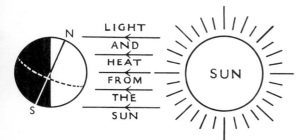

From March to September the northern hemisphere is tilted towards the sun

Christmas on Bondi beach, Sydney, Australia

195

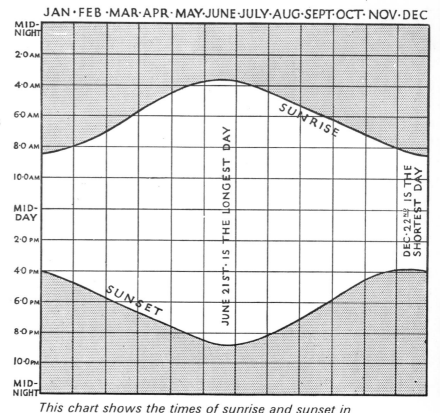

Because the earth is tilted, the days lengthen as one travels north from the equator in June. At the North Pole in summer there is no darkness, even at night. Even in the north of Scotland the sky is never really dark in June. But the hours of darkness and daylight together add up to twenty-four hours at all times of the year, as this chart shows.

This chart shows the times of sunrise and sunset in Manchester in each month of the year. The shaded part shows the hours of darkness

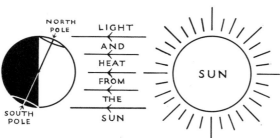

At the north pole in summer there is no darkness, even at night

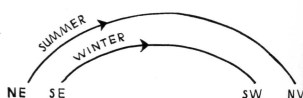

The path of the sun in summer and winter

The sun in summer and winter

In *summer* the sun rises in the north-east and sinks in the north-west. At mid-day it is high in the sky, and casts short shadows.

On a sunny day in the middle of *winter* the sun is never very high in the sky, and so it casts long shadows. In winter, the sun rises in the south-east and sets in the south-west.

196

17 Britain's weather

In Northern Germany in the middle of winter it is bitterly cold. Anyone going out-of-doors wears fur-lined boots and a fur hat, as well as gloves and plenty of warm clothing. In Chicago, and in New York too, it is so hot in mid-summer that everyone who can leaves the city.

Although Britain is about the same distance from the equator as these places, its weather is never quite so hot or so cold. Why is this? It is because Britain is surrounded by sea, which gets warm slowly and loses its heat slowly. As a result the winds which blow from the sea are never very hot or very cold.

Nor does Britain have a definite wet or dry season, as do the countries of Southern Europe which have nearly all their rain in winter, and long dry summers. Britain's mild weather, with rain at all times of the year, is usually called *temperate* or *maritime* (affected by the sea).

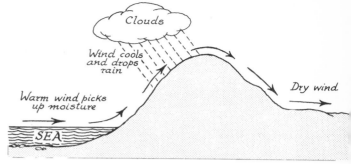

Why the south-west wind brings rain to the west of Britain

In winter the west of Britain has milder, damper weather than the rest of the country.

The wind which blows most often over Britain is a south-west wind. It is a warm wind in winter because it has blown from warm seas nearer the equator, bringing with it a warm sea current called the Gulf Stream. This current warms the seas round Britain.

After a shower on a hot summer's day the warm air dries a wet road very quickly. The water on the road becomes water vapour which is carried by the air. In the same way the south-west wind, blowing over warm seas, picks up moisture.

When the air cools, the water vapour turns back into drops of water. (On a cold day notice how water vapour from a hot bath turns into drops of water again when it meets the cold window pane.)

When the south-west wind reaches Britain it is forced to rise over the land. As it rises it cools, so bringing rain or drizzle to the west of Britain.

Learning to ski in the Cairngorms, Scotland, where snow often falls and lingers in winter

Look at these two maps and notice that most rain falls in the west and that least rain falls in the east.

Rainfall is measured in a rain gauge, which shows how much rain has fallen in a certain time. Parts of the Lake District have as much as 300 cm of rain in a year. Parts of East Anglia have less than 50 cm in a year.

In winter, eastern Britain has drier and colder weather than the rest of Britain.
East winds blowing from central Europe are very cold winds, for they come from lands which are far from the sea, and which have very cold winters. The cold east winds are dry too, for they have blown over land, not over sea.

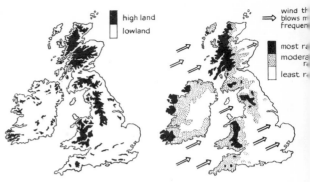

The high land of Britain is in the north and west. That is where there is most rain

In January the Scilly Isles usually have a temperature of 47° F (8° C) while East Anglia has less than 39° F (4° C).

In summer the south of England is warmer than the north of Scotland, because the sun is more nearly overhead in the south. In July the south-west of England has a temperature of 62° F (17° C) while in the north of Scotland it is 55° F (13° C). Farmers in Devon, Cornwall and the Scilly Isles grow early fruit and vegetables and sell them in cooler parts of Britain where the crops ripen later.

Many people go south for their holidays. In summer the south coast is usually the warmest part of Britain. East winds are warm winds in summer, because they are blowing from countries which have hot summer weather.

Village cricket in Staffordshire during an anticyclone in summer

Although in Britain the weather is usually warm in summer and cool in winter, we cannot forecast the weather for many days in advance. The long range weather forecasts on radio and television are more often wrong than the forecasts for the next day. One day it may be windy, the next day hot, and the next rainy. This is because over Britain warm air from the equator meets cold polar air. Together these two currents of air cause unsettled weather. The warm wet air is forced to rise over the cold air, and as it does so the water vapour in the air turns to rain. The barometer "falls" and we know that there is a *depression* over Britain. Then winds are often strong and there may be storms.

Sometimes in summer the area of depressions moves north, over Iceland and Norway. Then Britain has long periods of fine weather, with clear skies and very little wind. The barometer "rises" and we may even have a "heat wave". We say that there is an *anticyclone* over Britain.

July and August are two of the wettest summer months

199

During an anticyclone in winter it is usually cold and frosty at night, with clear skies. In November, an anticyclone may bring mist and fog.

18 A dairy farm in Cheshire

Mr. Gray has a small dairy farm in Cheshire. He keeps thirty cows and sells the milk which they give.

The Cheshire Plain is a good district for dairy farming. The soil is rich and heavy, and there is enough rain to make the grass grow well. Mr. Gray's fields are used for grazing by the cows and for growing hay or other cattle fodder.

For seven months of the year, from April to October, the cows are out in the fields day and night. But in November, when the nights are colder, they are brought into the cowshed at night. In the coldest part of the winter, from January to March, the cows are kept in the yard during the day and in the cowshed at night.

While the cows are in the yard, they are fed on hay, kale and grass which has been stored. Mr. Gray also gives them "cattle cake" (which is made from cotton seeds, amongst other things).

Other animals on the farm

Mr. Gray finds it useful to keep some sheep and pigs. Cows in the fields eat by pulling off the long grass with their tongues; they cannot eat the short grass. So in winter, when the cows are in the yard, Mr. Gray fattens sheep on the pastures.

They can nibble all the short grass which the cows have left, and they manure the fields at the same time. Pigs are also useful, because they grow fast and give a quick return of cash to the farmer. Any skimmed milk (from which the cream is removed) is fed to the pigs.

Cows feeding on kale, controlled by an electric fence

A milking parlour. Milk from the cows is pumped through a pipeline into a refrigerated tank

When a young cow, called a heifer, has her first calf, she begins to produce milk to feed it. But the calf can be taught to drink milk from a bucket, and later it can be fed on soft hay, meal, water and skim milk.

Soon the cow can join the milking herd. She is milked twice daily, once in the early morning and again in the late afternoon. Soon she is giving 18 litres of milk a day—enough for nearly 100 bottles or cartons of school milk!

Before Mr. Gray begins milking he washes his hands and puts on a white coat. Then he sponges the cows' udders before he fits the cups of the milking machines on to them. The milk is carried by pipeline into a tank where it is cooled to prevent germs increasing. On its way to the tank the milk is weighed so that Mr. Gray knows how much milk each cow is giving. Then he can give it the right amount of food.

From farm to home
Mr. Gray has about 450 litres of milk to sell every day. (Not all the cows are "in milk" at the same time.) A few farmers bottle their own milk and sell it on a local milk "round", but the milk from Mr. Gray's cows is bought by the Milk Marketing Board, for sale in the towns in South Lancashire.

A bulk tanker collecting milk from the farm tank

As Mr. Gray's milk is stored in a tank instead of in the traditional churns, the milk is collected by a bulk tanker. This makes the handling easier and quicker and allows the milk to be kept cool.

At the bottling plant the milk is heated to kill any germs—this is called *pasteurising*. Then it flows along a pipe to the bottling machine. From the time the milk is taken from the cow, to the time it reaches our homes, every care is taken to make sure that it is pure and clean.

Nearly three-quarters of the milk produced in Britain is sold as liquid milk. Of the rest, some is made into powder for baby food; some is sold to chocolate manufacturers for making "milk chocolate"; some is skimmed, and the cream is sold separately as cream or as butter. Some of the milk is made into cheese, or tinned or dried.

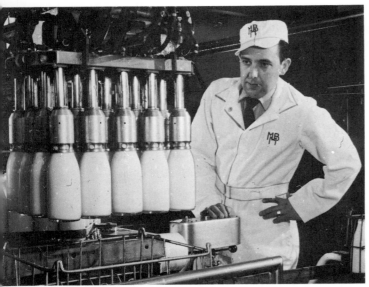

As well as the Cheshire Plain, there are many other parts of Britain which are good for dairy farming, including the south-west of Scotland, Wiltshire and Somerset.

Twenty full bottles of milk are automatically lifted into a crate

19 A fruit farm in Kent

London is one of the biggest built-up areas in the world: over eight million people live there. Think how much food they need even for one day! No wonder the farmers and market-gardeners around London, in the "Home Counties", grow as much fruit and vegetables as possible, to sell to the people of London and other large towns.

Mr. Cook is one of the farmers who send their crops to London. His farm is in the Weald, a district of fine farmland between two ranges of chalk hills, the North Downs and the South Downs. In the Weald the soil is a rich loam (a mixture of sand and clay) which is very good for farming.

Mr. Cook's farm covers 70 hectares of land. (A hectare is 10,000 square metres, the size of a square with sides 100 metres long.)

A knife and secateurs, using in pruning

Cherry blossom time in Kent. The tall buildings are oast-houses, in which hops are dried

Mr. Cook's farm

This is how the farm is planned:

1 *orchards* of apples, pears, plums, cherries:
50 hectares

2 *hop gardens* (the hops are used to flavour beer): 8 hectares

3 *arable land* (land which grows crops to feed the animals): 4 hectares

4 *pasture* for the sheep and bullocks, which provide valuable manure for the orchards: 8 hectares

Cold air flows down slope

No trees in frost 'pool'

Trees grown on a slope are not harmed by frost, which flows down the slope

Blanket of smoke

A blanket of smoke keeps in the warm air

Nearly all Mr. Cook's land slopes slightly to the south, facing the sun. This makes it warmer during the day and helps the fruit to ripen. The slope also helps to keep the trees free from frost, which is very harmful when the trees are in blossom.

How Mr. Cook looks after his trees
If there is danger of frost when the blossom is out, before the fruit has "set", Mr. Cook burns thick oil in the orchards. The smoke from the oil forms a "blanket" over the trees, and protects them by keeping in the warm air near the ground.

The fruit trees are the most important part of Mr. Cook's farm and he takes very good care of them. The orchards are regularly manured with chemicals, and by allowing sheep to graze between the trees.

If the fruit trees were allowed to grow naturally they would become so tall, and the branches so overcrowded, that the fruit would be small and difficult to pick. So in winter, while the trees are "sleeping", Mr. Cook trims some of the branches and thins out the shoots. This is called *pruning*.

Pests and diseases must be controlled all the year round, or the trees will be harmed by fungus diseases (such as brown rot and apple scab) and by greenflies, red spiders, codling moths and apple sawflies. To protect his trees Mr. Cook sprays them, about ten times each year, with insecticides containing nicotine and derris, and with fungicides which contain sulphur or mercury.

Spraying insecticide from a tractor

204

Picking apples

All the year round there is plenty of work to be done in the fields. One of the most important jobs is growing food for the animals, and the animals themselves must be cared for and fed.

Fruit trees do not last for ever, and occasionally an old orchard is dug out and replaced by new trees. These young trees are bought from a nurseryman who specialises in growing the latest varieties, which produce more fruit and less wood than old trees.

The trees are always planted in straight rows, to give their roots plenty of room to grow, and so that there is room for a tractor to drive between them, pulling spraying machinery, or a cart.

Picking the fruit

When the fruit is ripe, and ready for picking, Mr. Cook needs extra workers. Fortunately the different kinds of fruit are not all ready at the same time. The cherries are ripe in June, the plums in August, and the apples and pears in September and October.

The fruit is picked very carefully so that it is not bruised. Some of the apples are graded into sizes, put into boxes, and sent immediately to the market. But so that all the apples of Kent do not come on to the market at the same time, most of them are kept on the farms in gas storage chambers, where they will keep fresh until they are needed. Mr. Cook sends most of his fruit by lorry to the London wholesale fruit and vegetable markets.

The apples are graded into sizes and carefully packed

The man in the "crow's nest" cuts down the hop plants

Hops

The hop plants grow on a trellis-work of poles and wire, and the hop gardens have tall hedges round them so that the hops are not blown down by a strong wind. The hops form in late July, ripen during August and are ready for picking in September.

Years ago the hops were picked by families who came down from London. Now a team of men with a tractor cut down the hops and take them to the machine which picks or strips the hops. Then they are dried in sheds heated by oil burners.

Other counties where fruit is grown

Nearly all Britain's hops are grown in Kent, but fruit is grown in other counties where there is good soil and enough sunshine for ripening. There are fine orchards in Worcester, Hereford, Devon and Somerset. The Vale of Evesham is famous for growing fine plums. Strawberries, raspberries, currants and gooseberries (which are called *soft fruit*) are grown in Sussex, Essex, Hampshire and parts of Perthshire and Angus in Scotland. Because soft fruit does not keep, much of it is tinned or made into jams.

Some kinds of fruit, such as bananas, oranges and pineapples, cannot be grown in this country, and so we have to import them.

Hops

Blackcurrants

Gooseberries

Strawberries

206

Working in a glasshouse, where hot water pipes keep the air warm. Lettuces and tomatoes are being picked, and a cultivator is preparing the soil for the next crop

20 A market garden in Bedfordshire

Mr. Simpson's market garden is in Bedfordshire. It is really a small farm growing fruit, vegetables and flowers on every inch of its rich, heavily manured soil.

Part of Mr. Simpson's land is covered by glasshouses which are heated by hot-water pipes. The plants which he grows in the glasshouses are watered regularly and can be kept much warmer than those out of doors, and so they grow well and ripen earlier.

Tomatoes

To have plenty of tomatoes in the summer, Mr. Simpson must sow the seeds early in December. He sows them in shallow boxes and puts them in a hot glasshouse to germinate.

Eight weeks later the plants are planted out in a cooler glasshouse and tied to wires, so that they do not fall over. They are watered regularly, and given a special fertiliser. Mr. Simpson's men take out the side shoots so that each plant grows only eight or ten trusses (or bunches) of good-sized tomatoes and reaches a height of 2–2·5 metres.

The tomatoes are picked when they are just turning orange-red. They ripen quickly: in fact they will probably be red by the time the boxes arrive at the greengrocer's shop.

A tomato plant. The tomatoes at the bottom ripen first

207

Taking packets of quick-frozen vegetables from a plate-froster. Some of the vegetables come from market gardens, but most are specially grown by farmers

Dwarf beans

Dwarf beans grow well in Mr. Simpson's soil. The seeds are sown during May, in shallow furrows. As the plants grow they are weeded, and thinned. By mid-July the first beans are ready for picking, and more are ready every few days for several weeks until the frost comes.

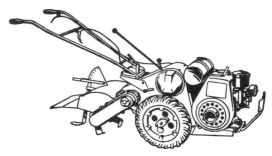

This rotary cultivator has a 5 hp engine. As the blades spin, they break up the soil

Flowers

Flower-growing is an important part of the market gardener's business. He grows flowers which will travel well, without drooping, and which he knows are popular —freesias, carnations, daffodils.

Lettuce

Mr. Simpson grows winter lettuce in the glasshouse before the tomato plants go in. Later, he raises outdoor lettuces in rich soil where they will "heart up" well. In hot weather the lettuces are sprayed with water from pipes which are moved about the fields.

Selling the produce

Mr. Simpson telephones the markets to find out in which town he can best sell his crops. Sometimes he is lucky, and just when there is a shortage at the market, he has a whole glasshouse full of tomatoes ready for picking. Sometimes he is unlucky, and he gets a poor price for his tomatoes.

Some market gardeners grow their crops especially for sale to a canning factory, or to a quick-freeze factory.

Some market gardeners grow fruit and vegetables for canning

208

At markets such as this one at Liverpool, fruit and vegetables are sold by the wholesale merchants to the retailers

Crops and flowers must be packed very carefully for market. Many growers now use plastic bags, and sometimes vegetables are washed and scraped ready for the saucepan.

Mr. Simpson employs many skilled workers and so his weekly wage bill is high. At the busiest times of the year many of his vegetables are picked by "piece workers" who move from farm to farm, and are paid according to the amount they pick. For this reason they work very quickly and expertly and it pays the farmer to employ them.

Most market gardens are near to large towns, in places where the soil is good: in north Kent, in Lincolnshire, Middlesex, Lancashire and Worcestershire, as well as in Bedfordshire.

These lists show some crops grown by market gardeners. Many of these crops are also grown by farmers, and on allotments and gardens.

Quick and early crops	Crops which take longer to grow	Root crops which can be stored	Soft fruit	Flowers
Lettuces Radishes Peas Dwarf beans	Spring onions Cauliflowers Brussels sprouts Cabbages	Carrots Potatoes Parsnips Turnips	Gooseberries Raspberries Strawberries Blackcurrants	Daffodils Carnations Gladioli Chrysanthemums

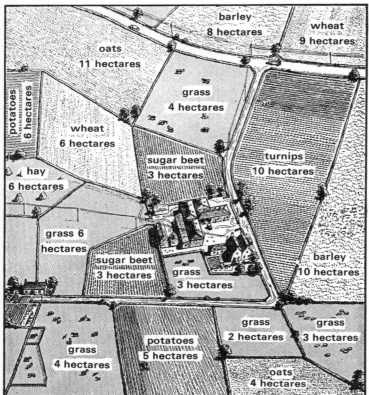

A plan of Mr Wilson's farm

21 A mixed farm in Eas Lothian

Mr. Wilson's farm of 120 hectares is in th county of East Lothian in Scotland. Th land is flat and easy to plough, but in thi part of Britain the growing season for crop is fairly short, and the weather is never ver hot.

Mr. Wilson grows crops in most of hi fields, but he keeps livestock as well, and s his farm is called a "mixed" farm.

Planning the crops

If Mr. Wilson grew only one crop, hi workers would be very busy at seedtim and harvest and would have little to do fo the rest of the year. But by planning th crops, the work of the farm can go o steadily all through the year.

Mr. Wilson does not grow the same cro in the same field every year. Grain crop such as wheat, barley and oats quickly tak the goodness out of the soil. So after grain crop he grows grass and clover, o swedes. These are good for the soil an also make good fodder for the animals.

A six-year rotation

Changing the crops in a field is calle "rotation". Mr. Wilson has chosen a six year rotation. The list at the top of the nex page shows what happens in six of Mr Wilson's fields during one rotation period

wheat
bread

oats
porridge

barley
beer

A Six-Year Rotation	11-hectare field	6-hectare field	9-hectare field	10-hectare field	8-hectare field	4-hectare field
1970	oats	potatoes	wheat	roots	barley	grass
1971	potatoes	wheat	roots	barley	grass	oats
1972	wheat	roots	barley	grass	oats	potatoes
1973	roots	barley	grass	oats	potatoes	wheat
1974	barley	grass	oats	potatoes	wheat	roots
1975	grass	oats	potatoes	wheat	roots	barley

The crops

Mr. Wilson is lucky to have a good loam soil on his farm: it is rich enough to grow grain and "stiff" enough for heavy-eared crops like wheat and barley to root well.

Mr. Wilson is growing oats in one of his fields this year. Let us see what happens in this field during the next six years.

A tractor pulling a seed drill

First year: Oats

Mr. Wilson sows the oats in March, and harvests them in August. Some of the oats go to the miller to be made into porridge oats, but most are needed as fodder for the sheep and cows during the winter. The straw, too, makes good fodder.

Oats are grown on most farms in Scotland, for they do not need as much sunshine as wheat or barley, and grow well even where there is much rain.

Harvesting beet. The machine lifts the beet, shakes off the soil, "tops" the beet, and loads them on to a cart

211

The potatoes are put into paper sacks and stored in sheds

2nd year: Potatoes

Mr. Wilson ploughs the field twice to prepare it for planting the potatoes in the spring. They grow best in a loose well-ploughed soil. Some potatoes are sold for eating, some are sold for seed. The smallest potatoes are fed to the pigs.

Sowing potatoes. The seed potatoes fall down the two tubes and are covered by the ploughshares

3rd year: Wheat

In November the land is ploughed, manured and sown with wheat. The wheat is ripe for harvesting in the following August. It is sold to a miller to be ground into flour for making bread, biscuits and cereals. Wheat grows best in a heavy soil with plenty of sunshine.

This transport box can be fitted to a tractor. It holds loads of up to 350 kg

4th year: A root crop

The next year Mr. Wilson grows root crops, such as turnips, swedes or sugar beet. The soil is ploughed and harrowed to break it up finely. While the root crop is growing it is kept free from weeds, and thinned, or *singled*, so that only good well-spaced plants are left.

Most of the turnips and swedes are dug up as they are needed, and fed to the animals. The rest are stored in clamps. Sugar beet are dug up in November and sent to a sugar refinery.

5th year: Barley

During the winter the field is ploughed and harrowed again. In March it is sown with barley. When the young shoots are a few centimetres high Mr. Wilson sows grass and clover between the rows. These grow slowly, so that when the barley is cut, the grass and clover are only a few centimetres high. Then the sheep graze on the young grass. Some of the barley is fed to the animals. The rest is sold for making beer.

6th year: Grass and clover

The following spring the grass is cleared of stones and rolled. Then it is left to grow long and in June it is cut for hay. The sheep graze on the second growth of grass and then the field is ploughed before Christmas, ready to start the six-year rotation again with a crop of oats.

Mr Wilson sows seed to grow his grass and clover

Mr Wilson's animals

Mr. Wilson keeps a flock of fifty sheep. He fattens all the lambs he can on grass and sells them when they are 4–6 months old. Those not fattened by then are "folded" on fields of turnips and swedes and sold during the winter. Nowadays most people like small joints of meat and they prefer young lamb fed only on grass.

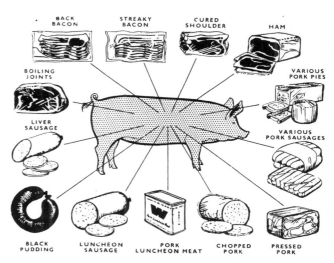

Mr. Wilson buys his pigs when they are only a few weeks old. They are fed on barley and oats.

Twenty Friesian cows are kept for their milk. They are out in the field from May to October, but they spend the winter in a sheltered strawyard. During the winter they are fed on hay or silage, "roots" and dairy meal.

Mr. Wilson is just one of thousands of farmers in Britain who have mixed farms. Not all these farmers keep the same livestock or grow the same crops. But all of them try to make a living by growing crops to sell and to feed to the animals, and by keeping livestock which provide meat, milk—and manure.

22 Other farms

Sheep farming

Sheep are very useful for they give us wool
as well as meat, and even on poor land they
can find enough to eat. Many sheep graze
on the limestone hills of southern England,
especially in Hampshire, where the grass is
short and wiry. Hardier breeds are kept on
hill and moor where the soil is thin and the
weather is often cold. There are many sheep
in Scotland, Wales, the Pennines and the
Lake District.

Sheep in folds, feeding on kale

Beef production

The best beef comes from pure beef cattle such as the Aberdeen Angus, Hereford or Devon.
They make really juicy steaks—if you can afford them! But most of us eat beef from dairy
cattle, such as the Friesian.

The roast beef of old England used to be famous but nowadays people eat less beef than a
few years ago, and more pork, chicken and tinned meats.

Poultry and pigs

Nearly everyone eats chicken today. The birds are kept in very large numbers in special
buildings. Many farmers, particularly in East Anglia and the Lothians, specialise in keeping
pigs because they can easily grow enough barley to feed them.

Arable farming

Arable farms grow the crops which we eat:
wheat for bread, oats for porridge, barley,
potatoes, etc. Most arable farms are on the
flat lowlands of eastern England and
Scotland, where the soil is good and there
is less rain than in the west. In north-east
Scotland oats and turnips are the main
crops. Further south, where the weather is
milder, crops such as wheat, barley, sugar
beet and potatoes are grown.

Feeding pigs with barley meal

23 Fishing

Britain is surrounded by shallow seas. In these seas fish thrive, because there is plenty of food for them to eat. They live on smaller fish, or on plankton, which consists of tiny sea creatures and plants.

This map shows the main fishing grounds in the seas around Britain. But the fish do not stay in one place all the time. They move about during the year, and the skipper of a fishing boat has to judge where the fish are likely to be. Often the skipper uses radar to help him to find the fish.

Some of the fish, such as cod, haddock, plaice and sole, live near the bottom of the sea, where they find their food. Other fish, such as herring and mackerel, live and feed near the surface of the sea.

Fishing grounds in the seas around Britain

Trawler fishing

The fish which swim near the bottom of the sea are usually fished for by boats called *trawlers*. These boats pull a net, shaped like a bag, along the bed of the sea. Once the fish are in the net they cannot escape. Some trawlers fish in the North Sea and to the north-west of Britain. But most trawlers sail for many days to reach their fishing grounds: eastwards to Bear Island and the Barents Sea or westwards to Greenland and Newfoundland.

The trawlers meet icy seas and stormy weather and are often at sea for three weeks at a time. The fish caught on one trip may be worth as much as £10 000. The fish are packed in ice to keep them as fresh as possible. Freezer trawlers freeze the whole of their catch at sea within a short time of catching, so that the fish is really fresh when it reaches port.

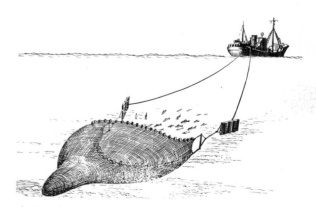

A trawler pulls its net along the sea bed

215

Baiting the hooks on a "great-liner"

Fishing with lines

Some of the bigger fish swimming near the sea bed, such as cod and halibut, are caught by boats called "*great-liners*". These boats put out long lines with hundreds of hooks on them. The hooks are baited with pieces of fish. When the line is lifted the cod and halibut are taken off and the hooks are re-baited. "Great-liners" fish for halibut on the rocky sea bed, in places where a trawl net cannot be used because it would be damaged.

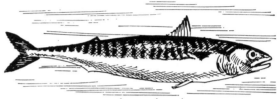

A mackerel

A box of herring

Drifters

Fish which swim near the surface are caught by *drifters*, and by *seiners*. The drifters put out nets which hang in the water like big tennis nets. The top of each net is held up by coloured buoys, and the bottom is kept down by a heavy rope called a *leader*. One boat may put out a "fleet" of as many as eighty nets. The nets have a mesh which is wide enough to let the small fish through. But as the larger fish try to swim through the net they catch their gills in it and cannot escape. Herring and mackerel are caught by drifters.

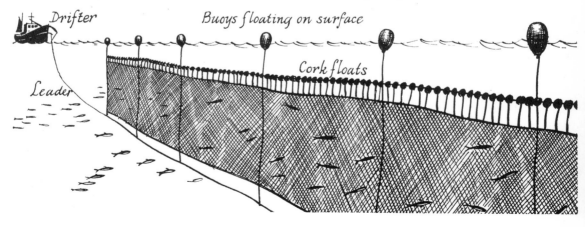

Drifter Buoys floating on surface Cork floats Leader

THE NORTH SEA

The colour picture overleaf shows some of the things which are being done in and around the North Sea. As well as fishing, which is described in this chapter, men are searching for natural gas and oil, they are digging for coal, manning the lifeboats and lightships and enjoying their holidays.

Natural gas 1 Drilling rigs are floated out and anchored. Each rig has a crew of about 50 men, who work 12-hour shifts for days at a time. One third of the crew is always on leave.

2 When there is sandstone deep under the earth's surface, gas, oil and water soak into the pores of the stone. The rock above presses on the gas, so that when a hole is drilled the gas is squeezed out by its own pressure.

3 Oil is often found mixed with gas and water.

4 The well may be 900 mm in diameter at the top and 200 mm in diameter 3000 metres down.

5 The pipeline is 3 m under the sea bed. It is covered with steel and concrete.

6 The first North Sea gas came ashore at Easington, north of Spurn Head. Natural gas gives great heat. It has no smell so a smell is added for safety.

7 Helicopters carry men and supplies to the rigs.

Coal 8 The National Coal Board's boring tower showed that 550 million tons of coal lie off the Durham coast.

9 Coastal collieries often extend their workings under the sea.

continued opposite page 217

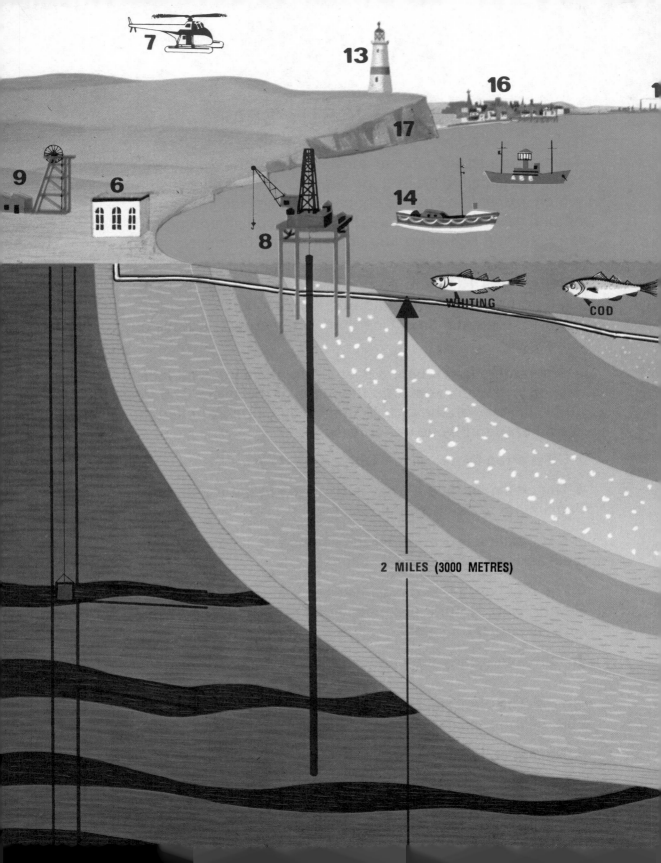

7

13

16

17

9

6

8

14

WHITING

COD

2 MILES (3000 METRES)

THE NORTH SEA

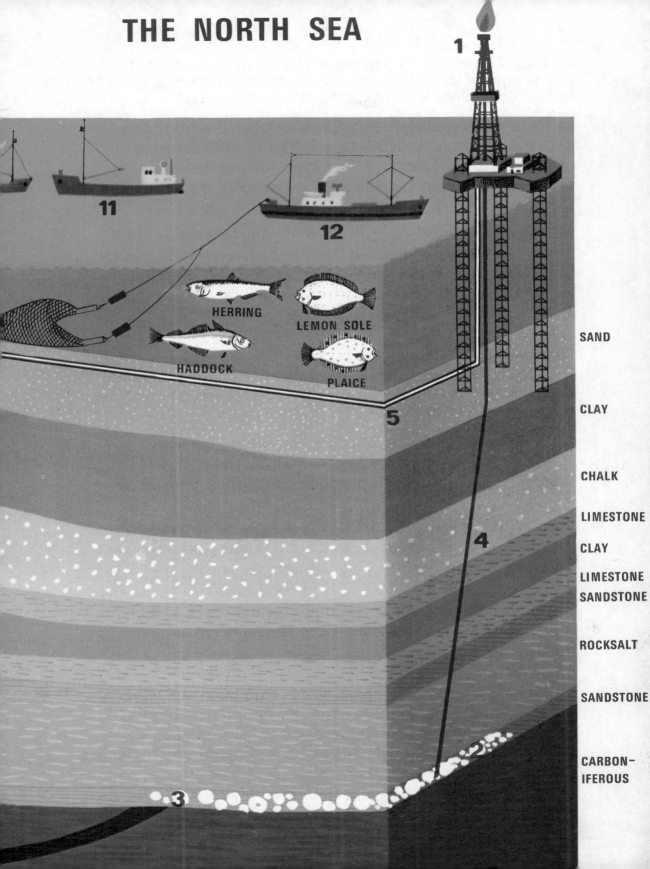

1

11

12

HERRING

LEMON SOLE

HADDOCK

PLAICE

5

4

3

SAND

CLAY

CHALK

LIMESTONE

CLAY

LIMESTONE

SANDSTONE

ROCKSALT

SANDSTONE

CARBON-
IFEROUS

Fishing 10 Inshore fishing boats (12-25 m long) rarely go out of sight of land. They catch crabs, lobsters and many kinds of fish, using lines and nets and pots.

11 Near-water trawlers (25-35 m long) fish the North Sea. Middle-water trawlers (35-45 m long) fish the Faroes and off Iceland. Both catch cod, haddock, whiting, hake, turbot, plaice and other flatfish. The crew is 7–9 men, the average trip is 10 days.

12 Distant-water vessels (up to 75 m long : minimum crew 20) sail mainly from Hull and Grimsby to Norway, Iceland and Greenland. They catch mostly haddock and cod. The average trip is three weeks, but factory trawlers can be out for three months.

Safety at Sea 13 Seventeen lighthouses and 22 red-painted lightships in the North Sea come under the supervision of Trinity House ; 8 of the ships send weather observations to the Meteorological Office.

14 There are 54 lifeboat stations on the North Sea, with 432 voluntary lifeboatmen, 8 to each boat. The Humber lifeboat is the only one in the country with a full-time crew.

Shipbuilding 15 In 1970, 156,000 men were employed in shipbuilding and ship-repair, of whom 34,000 worked on Tyneside.

Holidays 16 Twelve million people visit the East Coast on holiday each year. Scarborough, the largest resort, attracts 1 million people.

Erosion 17 Much of the East Coast is crumbling at a rate of 1-2 m a year. In parts of north Norfolk, around Flamborough Head, and on Bridlington Bay, erosion is much greater. Millions of pounds are being spent on reclamation and defences.

Based on material from *The Observer*

Herring swim in shoals, thousands of them together, so they are easily caught when they come to the surface to feed.

Purse seining

Drifting is a very old method of fishing. Nowadays more herring are caught in purse seine nets. When the captain finds a shoal, he "shoots" the net. The boat circles the shoal, trapping the fish in the net. At the bottom of the net there is a rope which pulls the net tight under the fish. The herring cannot escape, and they are lifted out of the sea with a dip-net.

Many of the purse seining boats are drifters like this one, fitted with the new kind of net, and modern equipment such as echo sounders and radar.

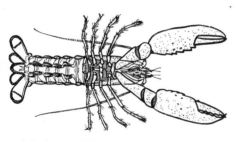

A lobster (the underside)

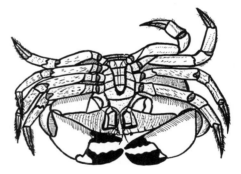

A crab (the underside)

A lobster pot, half made. The lobsters go in through the funnel shaped hole and cannot get out

Fishing near the coast

All round our coast there are many small fishing ports. From these ports small fishing boats go out almost every day. These "in-shore" boats rarely go out of sight of land or are away from home for more than two days. They use nets and lines to catch fish, and lobster pots to catch "shell fish", such as lobsters and crabs.

217

Herring being landed. .They are sucked through a pipe, and may be bought in boxes at the quayside. The men are looking at a catch which has just been landed

Landing and selling the fish

Most trawlers dock, and unload their fish, at Hull and Grimsby on the estuary of the River Humber, at Lowestoft in East Anglia, at Aberdeen in Scotland, or at Fleetwood in Lancashire. On the ships, the fish has been stored with ice to cool it.

At the port the fish is quickly sold, packed in boxes with ice to keep it fresh, and loaded into lorries, which carry the fish to markets in towns and cities all over Britain. Special fish trains go to the big cities, such as Birmingham and London, where there is a famous fish market called Billingsgate where wholesale fish merchants sell the fish to the fishmongers and to the fried-fish shops.

On this map only the first letters of the fishing ports are given. Can you name the ports?

218

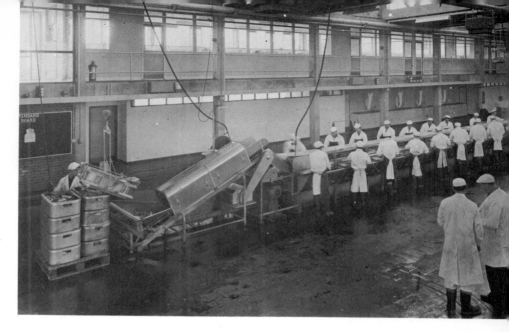

Packing freshly caught fish for quick freezing. In 1946 only 100 shops in Britain sold frozen foods. Now there are 130,000 shops, selling frozen food worth £125,000,000 every year

As well as the fishermen, there are many people whose work has a good deal to do with fish. How many can you think of?

Here are a few of them:
boat builders   lorry drivers
box makers   Billingsgate porters
canning and freezing factory workers
workers in fish and chip shops

Preserving fish

Sometimes, in winter, the seas are so stormy that the fishing boats cannot leave harbour. A few years ago this would have meant that there would be no fish in the shops.

But nowadays a good deal of the fish which is caught is "quick-frozen" and stored until it is needed. The factories preserve fish in many other ways. Some is canned; herrings are split, gutted, washed and hung in the smoke from burning wood-shavings to make kippers; some cod is salted and sent to South America and Southern Europe.

Even if the fish cannot be sold for eating, it is not wasted. It can be steamed to make fish meal, which has many uses, as you can see below.

FERTILIZER   DOG FOOD   FISH MEAL   POULTRY FOOD   PIG FOOD

Some of the many uses of fish meal

24 Transport and travel by road

Bert is a long-distance lorry driver. His depot is in London and he usually drives lorries which are carrying loads to Scotland or to the north of England. He drives them up the M1, the M6 or the A1 to a British Road Service depot or a transport café about 300 km from London.

WARNING SIGNS

Level crossing without gate or barrier ahead

Cross roads

Road works

Steep hill downwards

SIGNS WHICH GIVE ORDERS

Stop and Give Way

No cycling or moped-riding

Lorries prohibited

Some road signs

When he reaches the depot he hands the lorry over to another driver who will take it on to its destination.

Meanwhile Bert takes over a lorry which has come down from the north. He drives back to London where the lorry's load is delivered.

By changing lorries during his ten-hour shift Bert can return to his home base: he does not often need to find overnight lodgings. It is a hard, lonely life, but he is well paid and he enjoys driving.

When loaded, his lorry weighs about 25 tonnes. It is 9 metres long, and has a canvas cover to keep the load dry.

The journey from London

When Bert sets off, the roads are crowded with cars and vans. But when he reaches the motorway he knows that he should have a clear run until he reaches the hand-over point.

The Ross Spur Motorway (M50) connects the Bristol to Birmingham Motorway (M5) with South Wales

The journey between London and the north now takes much less time than it did a few years ago. Main roads have been improved and new roads built. Over half the journey from London to Glasgow can now be done on motorways.

There are hundreds of lorry drivers at work all over Britain, driving all kinds of lorries. The lorries carry goods quickly, without much handling, and this helps to keep down the cost of the things which they carry.

Britain's roads

The roads of Britain can be divided into four main kinds:

1 *Motorways.* These are fine modern roads made for fast, long-distance traffic. They are very safe, for they are wide and straight, and no roads cross them. One motorway (the M1) connects London with Birmingham, Doncaster and Leeds; another (M6) connects the Midlands with Lancashire and the road to Glasgow. Several other motorways have been built; some of them by-pass towns which were bottlenecks for traffic.

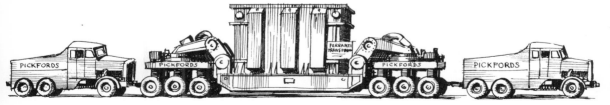

This Pickford's trailer has two tractors, one pulling and one pushing

Traffic in a jam in London

3. "B" roads are less important roads, linking towns and villages. Some of them are narrow and winding. They were good enough for the horses and carts which once travelled on them, but they are far too narrow for all the buses, lorries, vans and cars which travel on them today.

4. *Unclassified roads*—these are usually country lanes and the residential streets of towns. They have very little traffic apart from the cars of the householders and the vans of the tradesmen.

2. "A" roads are important roads, linking towns and cities. Some of them (the A1 for instance) have long stretches which are motorways. Others have long stretches which are almost as good as motorways.

A1 goes from London to Edinburgh
A2　　,,　　　　,,　　　　,,　　　　,, Dover
A3　　,,　　　　,,　　　　,,　　　　,, Portsmouth
A4　　,,　　　　,,　　　　,,　　　　,, Bristol
A5　　,,　　　　,,　　　　,,　　　　,, Holyhead
A6　　,,　　　　,,　　　　,,　　　　,, Carlisle
A7, A8 and A9 begin in Edinburgh

Many fine new roads and bridges have been built. But what will happen in ten years' time? Will there be so many cars and lorries that even the fine new roads will not be able to carry them? Should a law be made that "all heavy goods must be carried by rail", to leave the roads free for cars, buses and vans?

Cars and lorries use this tunnel under the Mersey

25 British Rail

Goods trains

A railway train is usually the best way of transporting heavy goods, for one goods train can carry the same load as fifty transport planes, or a hundred lorries. Train-loads of iron ore go from the docks to the steelworks. These trains need no sorting because all the wagons are going to the same place.

But many goods trains are made up of different wagons carrying all kinds of goods. When a wagon is loaded, it is put with other wagons going to the same part of the country.

The control room of a marshalling yard. You can see a marshalling yard from the air on page 145

Making up the goods trains

In every large town in Britain there is a railway goods yard. At the yard, wagons are sorted into trains by a shunting engine which pushes them from one line to another. Wagons going for long distances are taken to a large goods yard called a *marshalling yard*. There they are sorted again. An engine pushes the wagons over a hill or hump. As each wagon runs down the other side of the hump a man changes the points, so turning the wagon on to the right line. When the trains are complete, engines take them on their way. There are large marshalling yards at Toton (Nottingham), at March (Isle of Ely), Margam (Port Talbot), Kingmoor (Carlisle), Thornton (Fife), Tinsley (Sheffield), and at Crewe.

Freightliners

To save unloading goods from lorries into wagons and back into lorries, large containers are sometimes used. These can easily be lifted by a crane off the wagon and on to a lorry.

A freightliner terminal. The containers can be lifted off the wagon and on to a lorry

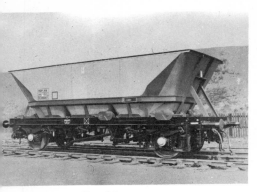

Coal hopper

*Petrol tank wagon,
holding
91,000 litres*

Steam-heated banana
wagon

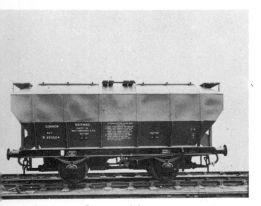

Covered hopper wagon

At present some goods trains travel at less than 50 km/h. This is because only the engine and guard's van have brakes, and it would be unsafe to travel any faster. But many goods wagons are fitted with automatic vacuum brakes, worked from the engine. Special goods trains carrying fish or milk to large towns are fitted with automatic brakes so that they can safely travel as fast as passenger trains.

Different wagons for different loads
Because goods trains carry so many different loads, there are many different kinds of wagon. More than half the wagons are open steel wagons for goods such as coal or iron ore which are not harmed by the weather. Goods which need protection from the rain go into covered wagons.

Passenger trains
The first railways were built to carry coal, but it was not long before passengers were carried. Nowadays there are over 16 000 passenger trains every day.

For a long journey a train is often the quickest and most comfortable form of travel. Trains are also useful because they carry so many people. About three million people travel by British Rail every day. Half of these people are workers in London, Glasgow, South Lancashire and Birmingham, who travel to and from work on the "rush hour" trains. Some trains hold as many as 1,000 passengers.

Suburban trains, which carry "rush hour" passengers, are designed to carry as many people as possible. Long-distance trains are more comfortable, with dining-cars where meals are served.

This car is being driven on to a train at Sutton Coldfield. It will be driven off when the train reaches Stirling, in Scotland

The main lines

Main lines link London with all parts of the country. There are two main routes from London to Scotland, one east of the Pennines via Peterborough, York and Newcastle, the other west of the Pennines via Crewe, Preston and Carlisle.

Trains for Holyhead and North Wales branch off the western route at Crewe. Trains to South Wales go via the Severn Tunnel. For Devon and Cornwall trains go via Taunton or Salisbury to Exeter, Plymouth and Penzance. South-east England is well served by frequent electric trains to and from London.

Several main lines link east and west Britain. There is a diesel service between Glasgow and Edinburgh. Fast trains run from Manchester to Liverpool and to Sheffield and through the gap made by the river Aire, from Leeds to Carlisle. But in most of Britain it is much easier to travel north and south than to travel east and west.

Railwaymen and railway towns

To keep the railways in good condition a large number of people are needed, working in the goods yards and stations, looking after the track, or working as signalmen, cleaners, porters or drivers.

At some places where the main lines meet there are large workshops where engines, carriages and wagons are built and repaired: Crewe, Swindon, Doncaster and Derby are some of these places.

Buffet car

A 3,500 hp 25 kv A/C electric locomotive, used for fast main-line trains

Modern locomotives

From the very first railways until only twenty years ago, coal-powered steam engines were the most common locomotives. But the steam locomotive has now been replaced by three other kinds of locomotive.

A *diesel locomotive* has a powerful oil engine, and is very useful for shunting.

A *diesel-electric locomotive* has an oil engine which drives a generator. This makes the electricity needed to drive the locomotive. Diesel-electric locomotives are used for pulling main-line trains.

An *electric locomotive* takes its power either from an overhead wire or from a third rail on the ground. It is useful for suburban lines, where the stations are close together, for it can start and stop very quickly. Larger electric locomotives are used for fast inter-city services.

Modernising the railways

The best plan would be to electrify all the main lines at once, but this would be far too costly. Instead some lines have been electrified, and diesel locomotives are being used on the others.

Railways are very expensive to run, and it costs a great deal of money to build new engines and stations, and to electrify the tracks. But even though there are roads in every part of the country, and more and more air services every year, a busy country like Britain needs good railways.

A diesel-electric locomotive

226

London airport

26 Air travel and transport

At London Airport

"BOAC announce the departure of their super VC10 Flight 501 to New York. Please proceed to Departure Gate 16. Thank you." You can hear announcements like this at London Airport, where aircraft take off for all parts of the world.

When passports have been checked the passengers are taken out to their airliner. They climb a short flight of steps into the aircraft. Each passenger settles into a comfortable, padded seat and buckles the seat belt round his waist. (This belt stops him being thrown forward if the plane jolts suddenly on the ground or in flight.)

Meanwhile the luggage has been loaded into place and the crew have checked that everything has been made ready for the flight.

A flight to New York

As the aircraft "taxis" from the airport buildings to the end of the runway the Captain receives his take-off clearance from the Control Tower. The wind direction determines which runway is used. He opens the throttles, and the plane moves down the runway, faster and faster, until suddenly the passengers realise that they are airborne.

The aircraft is fitted with a piece of equipment called an "automatic pilot". The Captain will switch this into operation during the flight and he or his First Officer will keep a check on its performance. The Engineer Officer is responsible for the smooth running of the four powerful Rolls-Royce jet engines.

A Super VC10 airliner, owned by British Overseas Airways Corporation (BOAC). Fully loaded it weighs 151 metric tonnes and carries 70 metric tonnes of fuel

The cabin staff, three stewards and three stewardesses, see that the passengers are comfortable in every way. They give them sweets on taking off, and serve meals which are warmed in a tiny galley, or kitchen.

Ten miles a minute

After seven and a half hours' flying the plane lands at New York. For most of the way it has been flying at nearly 1000 km/h at heights of up to 12 000 metres. The return flight, which is often helped by tail winds, takes about six and a half hours.

Britain's airports

Travel by plane is often more expensive than travel by train or by ship, but it is very much quicker and easier. Every week thousands of passengers fly from Britain's airports: businessmen who are in a hurry, and holiday-makers who want to spend as short a time as possible on the journey.

Britain's largest airports are London's Heathrow and Gatwick airports, Prestwick and Abbotsinch (Glasgow), Ringway (Manchester), Elmdon (Birmingham), Turnhouse (Edinburgh), Speke (Liverpool), Newcastle, and Belfast in Northern Ireland.

Aberdeen, Leeds, Luton, Cardiff, Bristol and Southampton have smaller airports, as do many other towns.

Serving a meal on board

228

Air traffic control

With so many planes flying over Britain it may seem surprising that planes hardly ever hit one another in mid-air. But no pilot can choose his own course. Instead an Air Traffic Control Centre tells him at what height, and by what route, he must fly. Every civil airliner is controlled all the time it is flying in an "airway" or "airlane". London is the control centre for Southern England, Preston for the north of England, Prestwick for Scotland and the eastern half of the North Atlantic Ocean.

Loading luggage into a VC10. Most airlines carry freight and mail as well as passengers

The airlines of the world

British European Airways (BEA) and British Overseas Airways Corporation (BOAC) are Britain's main airlines. BEA planes fly to nearly seventy different airports in Britain, Europe and North Africa. BOAC planes fly to the Middle and Far East, Australia and New Zealand, Africa, the Caribbean and North, Central and South America. All these are regular flights, keeping to a timetable. BOAC, BEA and other companies also transport cars for holiday-makers, and make special or "chartered" trips, carrying groups of people such as employees of British companies. Charter flights are usually cheaper than the regular airline fares.

Transport planes

Passenger planes carry mail and cargo, as well as passengers. Other planes carry only cargo. Goods which perish quickly, like tropical fruits, or cargo which would be harmed by a long sea voyage, like delicate machinery, race-horses, or animals for the zoo, are sent by air. All cargo is weighed and then carefully stowed so that it does not upset the balance of the plane.

A rally car being loaded on to an airliner

This is one way of dividing Britain into regions

27 The regions of Britain

If you look at a map of Britain in an atlas you will see that it is coloured. On some maps the counties are given various colours, so that you can easily see them. But on a physical map of Britain the land is coloured to show how high it is above sea-level. High land is brown and low land is green. In Britain, most of the high land is in the north and west. In the middle of England, and in the east and south, the land is much lower. Some of the low land has small hills, but all this land is less than 200 metres above sea-level.

The height of the land in the different parts of Britain helps to give each part its particular scenery. In some places there are flat plains. In other places there are wide river valleys, gently curving hills, high moorland, or rugged mountains.

According to the different kinds of scenery, we can divide Britain into different parts. But if we take into account other things as well, Britain can be divided into *regions*, each of which has its own character.

Here are some of the things which help to divide Britain into regions:

The height of the land—is it highland, upland or lowland?
The weather—how much rain falls? Is it a warm or a cold part of Britain?
The natural resources—can coal or iron ore be mined there?
The soil—is it good for growing crops?

1 *The Pass of Glencoe, in the Highlands of Scotland*

Scotland

1 The Highlands

Most of this part of Scotland is made up of very old hard rock, which has stood up to centuries of wind, frost, sun and rain. The mountains are craggy, the soil is poor and thin, and the rainfall is heavy. Farming is very difficult, but some hillsides have been planted with trees. In some valleys hydro-electric power stations have been built. In the east there is some lowland which is good for farming.

2 The Central Lowlands

Four-fifths of the people of Scotland live in this wide valley. The land is good for farming, coal is mined there, travel is easy, and there are deep sheltered estuaries. Many people who live in the Central Lowlands work in factories.

2 Building the Queen Elizabeth 2 on the River Clyde

231

3 Sheep on the Southern Uplands of Scotland

3 The Southern Uplands

This part of Scotland is not as high or as rugged as the Highlands. As on most high land in Britain, sheep are reared for mutton, and for their wool. On the lower land there are dairy farms. The Uplands join the Cheviots, a range of hills which divides Scotland from England.

England

4 The Lake District

This is a lovely area of rounded mountains, with forested slopes and beautiful lakes. Many of the people who live there earn a living by looking after the climbers, walkers and anglers who go to "the Lakes" for a holiday.

4 Walkers near Derwent Water in the Lake District

5 The Pennines

This range of hills forms a natural barrier between Yorkshire and Lancashire. The roads and railways follow the valleys through the hills, but even so it is still possible for travellers to be snowbound in winter. Streams from the Pennines provide soft water for the cotton mills of Lancashire, and the woollen mills of Yorkshire. The Pennines reach well down into England, as far as the Peak District of Derbyshire.

5 The Woodhead tunnels, under the Pennines

6 North-East England

Here there is a large coalfield, with many factories, mines, shipyards and docks.

7 Yorkshire

The Vale of York is a rich plain, where good crops are grown. In South Yorkshire there are many coal mines, steel works and factories.

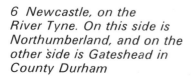

6 Newcastle, on the River Tyne. On this side is Northumberland, and on the other side is Gateshead in County Durham

233

8 Lancashire and Cheshire

In Lancashire there is a large coal-field. The busy ports of Liverpool and Manchester serve the factory towns on the River Mersey and the cotton towns north of Manchester. Market gardeners in south and west Lancashire grow vegetables for the people of the towns. Dairy cows are kept on the rich pastures of the Cheshire plain, and north Lancashire.

There is also an important chemical industry in Lancashire and Cheshire.

8 Two swing bridges on the Manchester Ship Canal. One carries a road, the other a canal

9 The Midlands

Most of the people live and work in the factory towns, particularly in and around Birmingham. In many parts of the Midlands there is good farmland where cattle are reared, sheep graze, and much fruit is grown. This is one of the most prosperous parts of Britain.

9 Stoke-on-Trent, a large town in the Midlands

234

10 Large machines can be used on the rich, flat land of East Anglia

10 East Anglia

Heavy crops of wheat and sugar beet are grown on the good flat farmland of East Anglia. Around the Wash is an area known as the Fens, much of which has been reclaimed from the sea. It has dark, rich soil. The Broads, a series of lakes joined by rivers, between Norwich and Yarmouth, are very popular with people who like boating holidays.

11 The West Country

The high moors, Exmoor, Dartmoor and Bodmin Moor, are bleak and almost uninhabited. On the lowland the weather is mild, and early vegetables and flowers are grown in the sheltered valleys and sent to London and other large towns. Round the coast there are many small fishing ports.

11 Picking daffodils on the Cornish coast

12 Southern England

Much of this part of England was once covered by forest. Today the only large forest is the New Forest, between the great port of Southampton and Bournemouth, a large holiday resort. Salisbury Plain is a large area of open downland where much wheat is grown. Sheep are kept on the uplands and dairy cattle on the meadows in the valleys.

12 The docks at Southampton

13 London

Dozens of towns and villages have grown and gradually joined together, to make the huge built-up area known as Greater London. Hardly any open space is left. Over eight million people live there, and the built-up area stretches for over 30 km from north to south and 50 km from east to west.

13 London. The Houses of Parliament, seen across the Thames at Westminster. Westminster Bridge is on the right

14 South-East England

Here there are two ridges of chalk hills, the North Downs and the South Downs, where many sheep graze. Between the Downs is the Weald, a district of fine farmland where fruit and hops are grown. Much food for the people of London comes from this region and many Londoners spend their holidays at seaside resorts around the coast.

14 Brighton, a seaside resort

15 North Wales

Sheep abound everywhere in the rugged country of North Wales. The rainfall is heavy and water from some of the lakes is piped to cities as far away as Birmingham and Liverpool. Snowdon is the highest mountain in either Wales or England. Many people from Lancashire and the Midlands go to North Wales for their holidays.

15 Llanberis Pàss, near Snowdon, North Wales

16 A valley in Merioneth, Central Wales

Wales

16 Central Wales
Few people live in Central Wales, and there are no large towns. Sheep graze on the hills. Aberystwyth is a holiday resort, a university town and a centre of Welsh culture.

17 South Wales
Here there are several long, narrow valleys, such as the Rhondda and Ebbw Vale, where coal and anthracite have been mined for many years. Today there are also tinplate works, and many new industries. In the towns on the coast there are large iron and steel works, and the ports of Swansea and Cardiff, which is the capital and business centre. To the east and west of the industrial region is good farming land which produces food and milk for the towns.

18 A factory in Northern Ireland

Northern Ireland

18 Northern Ireland
There are high blocks of mountains, with lowland between them and around Lough Neagh. Sheep graze on much of the highland while in the lowlands cattle, pigs and hens are reared and oats are grown. Most of the factories of Northern Ireland are in and around Belfast.

17 A large steel works at Margam, South Wales

Let's remember

Here are some facts and figures about Britain

The British Isles

The British Isles are a group of islands lying off the north-west coast of Europe. The largest island is Great Britain, made up of England, Scotland and Wales. The next largest island is Ireland, made up of Northern Ireland and the Republic of Ireland. There are many smaller islands.

The United Kingdom of Great Britain and Northern Ireland is made up of England, Scotland, Wales and Northern Ireland.

The high land

Most of the high land in Britain is north and west of a line drawn from the mouth of the River Exe, in the south-west, to the mouth of the River Tyne, in the north-east. Rainfall is heaviest in this high land.

Distance

From London to the most northerly point on the mainland of Scotland is over 800 km.

Natural resources

Britain's only natural resources are coal, natural gas, a little iron ore, a little water-power, and fish.

Power

Coal is still our main source of power, but gradually its place is being taken by oil, hydro-electricity and atomic power, and by natural gas.

Industry

The eight main industrial areas are: the Central Lowlands of Scotland, the north-east of England, South Lancashire, South Yorkshire, the Midlands, South Wales, London and Belfast. Of these eight areas, only two—London and Belfast—are not on a coalfield.

Farming

The highest land cannot be farmed. Sheep graze on the uplands. Crops are grown and cattle, pigs and poultry are kept on the lowlands.

Imports and exports

The farms of Britain can provide only half the food needed in Britain; the rest must be imported. Raw materials for the factories must also be imported. To pay for these imports, manufactured goods must be sold abroad.

People

Over fifty million people live in the United Kingdom. Four-fifths of them live in towns.

Weather

The weather is changeable. It is not too hot in summer and not too cold in winter. It may rain on any day of the year, but more rain falls in winter than in summer.

PART FOUR

Looking at
The World Today

The publishers are grateful to the following for permission to reproduce photographs:

Aerofilms Ltd 260, 263a, 321a, 322, 336a, 337, 351a; Australian News and Information Bureau 248, 249, 273a and b, 274a and b, 275a, 276a and b, 297a and b; Barnaby's Picture Library 247b, 265, 269b; Berne Lotschberg Simplon railway 314b, 315b; British Antarctic Survey 348; British Petroleum Co 247a; Camera Press 295, 299b; Central Office of Information 282, 284a, b and c, 286, 289; Chr. Salvesen and Co 350; David Gadsby 256; Erich Hartmann, Magnum 333a; Foto Film 317; French Government Tourist Office 310, 311a; French Railways Ltd 311b; Ghana High Commission 288; Government of India 267, 269a; Israel Government Tourist Office 291b; Italian State Tourist Office 315a; J. Allan Cash 294b, 296b, 327a and b; Japanese Embassy 308a, 309; John Hilleson Agency 270; Jonathan Rutland 290, 291a, 293, 294a; National Film Board of Canada 345a and b, 351b; New Zealand High Commission 250, 334, 335, 336b, 338, 339a and b; New Zealand Shipping Co 245; Novosti Press Agency 324, 327, 328a and b, 329a and b; Pakistan High Commission 268; Popperfoto 241, 244, 251, 252, 261a and b, 263b, 264a and b, 300a and b, 307, 308b, 312, 320, 330, 331, 333b, 340b, 341b, 349, 351c; Radio Times Hulton Picture Library 266, 299a; Ransomes and Rapier Ltd 298; Rob Wright 243, 283; Shell Photographic Unit 346; South African Information Service 301, 302, 303, 304; Swiss Cheese Union Inc 313b and c; Swiss National Tourist Office 313a, 314a; *The Times* 296a, 340a; United States Information Service 278, 279a and b, 280a and b, 319a and b, 312b, 323a and b, 341a, 342, 344

Acknowledgement is also due to the following for permission to reproduce maps and drawings:
British Broadcasting Corporation 252 (drawings), 318 (map); Controller of Her Majesty's Stationery Office, Crown Copyright 254, facing page 256; Odé, Paris, from *Le Japon* 308c; Sperry Gyroscope Co Ltd 253a; Twentieth Century Fund, New York from *World Population and Production* by W. S. and E. S. Woytinsky 342a
The maps are by Cyril Webber
Most of the drawings are by Geoffrey Whittam
The publishers are grateful to the New Zealand Shipping Co Ltd for their help in the preparation of chapter 1

242

CONTENTS OF PART 4

Two Indian girls

About part 4

This part of the book is divided into three parts:

THE HOT LANDS
THE TEMPERATE LANDS
THE COLD LANDS

In the first section, you will read about different kinds of HOT LANDS, from the tropical forests of Brazil to the plains of India and the deserts of Australia. In most hot lands the climate makes life difficult.

The second section is about TEMPERATE LANDS. In many of them the climate encourages people to work hard, so that in them we find the busiest nations of the world.

The third section describes COLD LANDS. In them, the climate is so harsh that people can provide themselves with food and shelter only by working tremendously hard.

When you are reading this part of the book, remember that some countries are so big that they cannot easily be labelled HOT, TEMPERATE or COLD. Huge countries such as Russia, India, Australia and the U.S.A. have many types of climate, peoples and occupations.

Above all, remember that CLIMATE chiefly decides how people live, work and amuse themselves.

1 A Voyage Round the World in a Cargo Ship

Here is the M.V.* *Tongariro*. She is a 20-knot refrigerated and general cargo ship of 8233 tons and she sails regularly between Britain (the "U.K." as sailors say) and Australia and New Zealand. She is a fine modern ship built in a famous ship-building yard on the north-east coast of England.

Loading the cargo

Long before the *Tongariro* begins a voyage, a plan of the ship is filled in to show exactly what cargo is to go into each hold. The holds are filled in such a way that the ship keeps on an even keel, and so that the cargo to be unloaded first is on top. All cargo is stowed with great care, so that it is not damaged, even in a rough sea.

The *Tongariro* loads the first part of her cargo in London docks. Cranes on the dock-side swing the cargo into the holds: cars, cotton goods, bags of cement, glass for windows, and earthenware pipes (which

* M.V. stands for motor vessel.

will be used for a new hydro-electric plant in Australia).

The *Tongariro* then sails on to Swansea and Newport in South Wales, to load tin plate and galvanised steel, and then to Liverpool to load machinery, cottons, woollens and chemicals. In fact many goods that have been manufactured from raw materials brought to the U.K. on a previous voyage of *Tongariro* and by other ships.

An electric crane

245

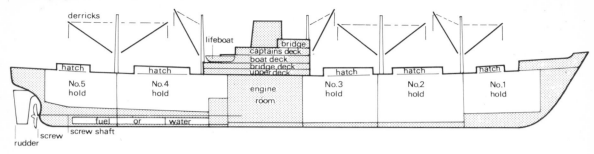

Plan of M.V. Tongariro

At last, all the cargo has been loaded and the hatches are sealed. The regular crew, who have been on leave, rejoin the ship at Liverpool, and the *Tongariro* sets sail for Australia. Stores and provisions are on board, the tanks of fuel oil for the engines are full, and every important piece of equipment on the ship has been carefully tested.

Liverpool to Australia

A Pilot steers the *Tongariro* out of the docks. He leaves the ship off Anglesey, and takes back to Liverpool the letters which the crew have written since the ship left port.

As soon as the ship is at sea, work begins on clearing away all the grime and dirt of loading. The decks and hatches are tidied and washed down, and ropes and gangways are stowed away.

The officers and crew settle down to the normal work of a long voyage. The officers keep watches (which really means "work shifts") on the bridge, in the radio room, or in the engine room. At sea, the Master of the ship is in command: his word is law.

The *Tongariro* sails down the west coast of Britain. On her way she sees lighthouses and lightships, which help to keep her on her course and guide her clear of dangerous rocks and currents. She passes many other ships: coasters, oil tankers, iron-ore carriers, and liners heading for Liverpool. Gradually there are fewer ships, until at last the *Tongariro* is out in the open sea and setting course for the Bay of Biscay. She sails past the Strait of Gibraltar, the western entrance to the Mediterranean Sea, en route to Las Palmas in the Canary Islands off the north-west coast of Africa where more fuel oil for the engines will be loaded in addition to fresh water for the engines and the

"Dropping" the pilot into his cutter

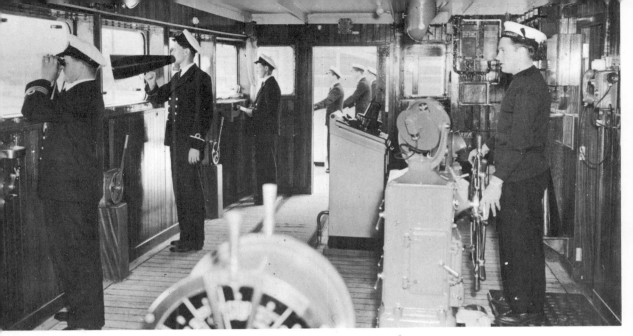

On the bridge

crew. Some modern ships, like *Tongariro*, have special equipment on board which enables them to make limited quantities of fresh water from sea water.

Whilst at Las Palmas, *Tongariro* takes on fresh vegetables and everyone on board receives their first letters from home since leaving Liverpool. *Tongariro* then sails for the long haul of 9500 miles (15 000 km) from Las Palmas to Fremantle, Australia.

Las Palmas

"Crossing the line"

"Crossing the Line"

About three days after leaving Las Palmas there is the ceremony of "crossing the line", when anyone who is crossing the equator for the first time is made a "citizen of King Neptune". If the ship has no swimming pool, a canvas tank is set up and filled with sea water. Then the new citizens are lathered, shaved, christened with such names as "Harassed Herring" or "Soulful Salmon" and tipped into the water.

The ship continues round the Cape of Good Hope with a brief stop of a few minutes about one mile off-shore to meet a launch sent from Cape Town harbour with more letters from home. The journey round the southern-most tip of Africa follows the route taken by sailing ships and early steamers on passage to Asia and Australasia before the Suez Canal was opened in 1869.

Discharging cargo in Australia

The *Tongariro* arrives at her first port of discharge: Fremantle in Australia. As soon as the ship is tied up, there is a surge of activity: gangways are brought up, cranes swing over the holds, and customs men and port officials hurry aboard. Australian dock workers, using both dockside cranes and the ship's winches, unload part of the cargo—motor cars and agricultural machinery—and put it into sheds on the wharf.

Fremantle, Australia's main port on the Indian Ocean. The harbour, which is at the mouth of the Swan river, is man-made

Some of the wharves near the main commercial centre of Sydney

When the cargo for Fremantle has been unloaded, the *Tongariro* sails past the big grain elevators where tramp ships are loading grain to take to Britain, and past a large oil refinery, out to sea. Three and a half days later she reaches Melbourne, where more of the cargo is discharged, and then she sails on to Sydney, the biggest and busiest port in Australia.

She sails under the famous Sydney Bridge and into the great landlocked harbour where ships are loading wool, skins, hides, tallow, eggs, canned fruit and canned meats.

At last the whole of the cargo which the *Tongariro* brought from the U.K. has been discharged. Then each hold is thoroughly washed and cleaned and the ship is ready to take on board refrigerated cargo in New Zealand.

Loading in New Zealand

In New Zealand, as in Australia, the crew feels really at home. They are in countries inhabited by people from Europe, many of them from Britain.

At Auckland, two holds are filled with mutton, lamb and butter to be kept at a temperature of 14°F (minus 10°C) and cheese at a temperature of 43°F (6°C).

Then the ship sails on to Napier. Here boxes of apples are taken on. They must be kept cool but not frozen. Great care is taken in loading and packing the cargo to make sure that things with a strong smell, such as cheese or wool, are not packed with apples or butter.

Looking across the harbour of Wellington, the capital of New Zealand

At Wellington, the capital of New Zealand, other kinds of food are loaded—tins of honey, mutton wrapped in mutton cloth, and beef wrapped in sacking.

Wellington has a beautiful harbour, and the town has fine modern buildings, including the New Zealand Houses of Parliament.

At Lyttelton, the port for Christchurch, more meat is taken on board, together with bales of wool. The *Tongariro* is now fully loaded and she sets course on her way home across the Pacific Ocean.

250

Unloading supplies on Pitcairn Island

The voyage home

The *Tongariro* sometimes calls at Pitcairn Island, midway between New Zealand and America. Many of the inhabitants are descendants of the crew of the ship *Bounty*, who took part in a famous mutiny in 1789. They paddle out in canoes and swarm all over the *Tongariro*, selling beads, as well as bananas, oranges and other fruit.

While the ship is at anchor, the crew "post" letters home. The letters are taken back to the U.K. in the *Tongariro*—but they bear the Pitcairn Island stamp.

When the *Tongariro* reaches Balboa she begins her journey through the Panama Canal, which links the Pacific Ocean with the Atlantic Ocean.

The canal is 50 miles (80 km) long; half of this is through an artificial lake, Lake Gatun, which was made by damming the River Chagres. Twelve locks, in three flights, carry ships up to and down from Lake Gatun, which is 85 feet (26 m) above

sea level, and small electric engines pull the ships through the locks.

Many United States ships, planes and soldiers are seen as the ship sails through the canal. On either side the land is covered with tropical vegetation. Wild bananas grow there, and monkeys swing from tree to tree.

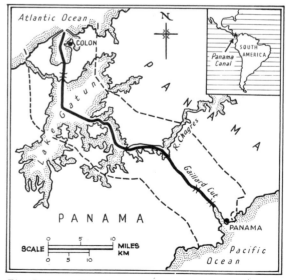

The Panama Canal

The Gaillard Cut on the Panama Canal

The Panama Canal

The journey through the canal takes about eight hours. Ships are allowed to sail during the day and night and set off at half-hourly intervals.

The Atlantic port is Colon which she leaves behind to take on fuel oil at Curacao, where it is cheaper than in the U.K., since it comes from the near-by oil fields of Colombia and Venezuela.

How a lock is used to raise a boat

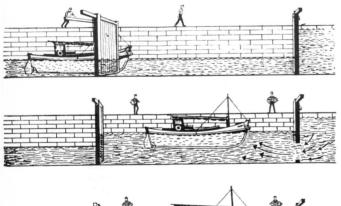

1 The lower lock gate is opened and the boat sails into the lock. The level of the water in the lock is the same as that in the canal below the lock

2 When the lower gate has been shut, the sluices in the upper gate are opened. Water passes into the lock and so raises the boat

3 When the level of water in the lock is the same as the level up-stream, no more water can flow

4 The upper lock gate is opened and the boat continues its journey

Home again

The *Tongariro* sails on across the Atlantic Ocean. When she is off Cornwall she sees the flashing light of Eddystone Lighthouse —a double flash every half minute. Each lighthouse flashes a different signal from those near to it, so that it can be quickly recognised.

Off Brixham, in Devon, a Channel Pilot comes aboard. He takes the ship as far as Gravesend where a River Thames Pilot takes over to steer the ship to the Royal Albert Dock. A Dock Pilot steers the ship, with the help of tugs, into the dock, where work begins at once to unload the refrigerated and general cargo. Most of the crew are "paid off" immediately, and they go on leave. As soon as the cargo is unloaded, the ship is cleaned and immediately the dockers begin to load the next cargo.

The log of the *Tongariro's* voyage

The voyage has taken 16 weeks, including time in port as well as at sea. Here is a log of the *Tongariro's* sailing time, not including time spent in the ports.

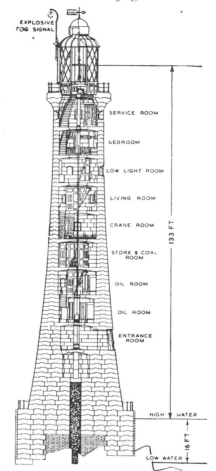

	Sailing time
Liverpool to Las Palmas	$3\frac{1}{2}$ days
Las Palmas to Cape Town	$9\frac{1}{2}$,,
Cape Town to Fremantle	$10\frac{1}{2}$,,
Fremantle to Melbourne	$3\frac{1}{2}$,,
Melbourne to Sydney	$1\frac{1}{2}$,,
Sydney to Auckland	$2\frac{1}{2}$,,
Auckland to Wellington	1 day
Wellington to Lyttelton	$\frac{1}{2}$,,
Lyttelton to Balboa	14 days
Panama Canal	$\frac{1}{2}$ day
Colon to Curacao	$1\frac{1}{2}$ days
Curacao to London	9 ,,
Sailing time	$57\frac{1}{2}$ days

Eddystone lighthouse

The Owers lightship, which is anchored off Selsey Bill, Sussex. Lighthouses and lightships are used as aids to navigation. Sometimes they mark places which are dangerous to shipping. At night they show a flashing light

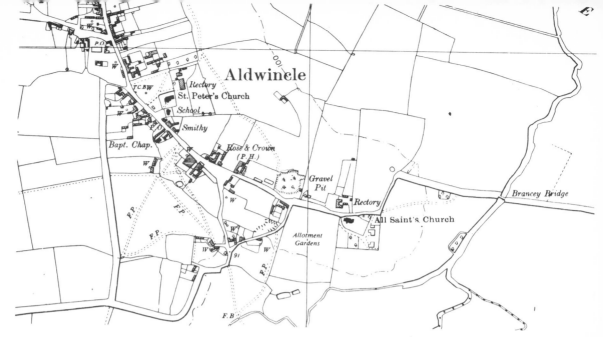

Aldwincle

A map of Aldwincle, Northamptonshire. (Scale 1:10 000—approximately 6 in to 1 mile)

2 Looking at Maps

What large-scale maps show

A map with a scale of 1:2500 is called a large-scale map. On such a map the shape of every building is shown, the houses are numbered, and even telephone kiosks are marked and named. Part of a 1:2500 map is shown below.

A map with a smaller scale (say 1:10 000) shows a larger area, and so cannot show the shape of every building. Rows of

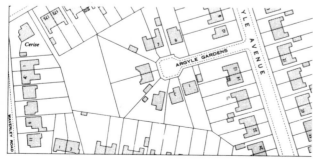

Part of a map with a scale: 1:2500

houses are shown, but not each separate house. On the 1:10 000 map above, initial letters have been used to mark such things as wells (w), a footpath (FP) and a Post Office (PO).

What a small-scale map shows

A map with a still smaller scale (such as 1:63 360, the scale of the popular "One-Inch" Ordnance Survey series where 1 inch represents 1 mile) clearly shows roads (red), rivers (blue) and woods (green). But the shape of many things such as Post Offices, railways and churches cannot be shown. Instead of showing the actual shapes of these things on the map, little signs, called *conventional signs*, are used. A One-Inch Ordnance Survey map is printed opposite page 256. Underneath the map is a key to some of the conventional signs used on it. You will see that it has a grid of faint grey lines all over it. Each square these lines make represents one square kilometre (1 km^2).

How to show high land and low land on a map

If you make a sketch map of your journey to school, you may want to show a place where you walk up a hill, or down a hill. You can do this by putting an arrow on a road, to show an uphill gradient, like this:

Or you can shade the high ground, like this:

But these methods are not very helpful, for they do not show clearly just where the ground is highest, or how steeply it rises.

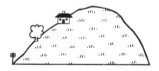

Here is a hill. Near the top is a cottage, and halfway up is a tree. At the foot is a telegraph pole. Let's see how we can draw this hill to show which parts of the hill are highest and which are lowest. First shade the picture of the hill to show all the land which is higher than the cottage:

Then shade all the land higher than the tree but lower than the cottage:

Lastly, shade the land between the telegraph pole and the tree:

Your drawing of the hill will look like this:

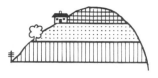

If we draw this hill, to show it as seen from above, and then shade the drawing, it looks like this:

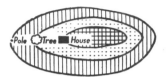

Lines have been drawn to separate the different heights on the hill. All the land on one side of these lines is higher than all the land on the other. The lines are called *contour lines*. They are used on maps to show how high the land is above sea level.

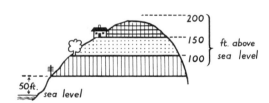

Suppose that the vertical distance between the telegraph pole and sea level is 50 ft. If the base of the tree is 50 ft higher above sea level than the foot of the telegraph pole, and the cottage 50 ft higher than the tree, then the contour lines drawn through these things are each 50 ft apart. We say that the *contour interval* is 50 ft.

On the One-Inch Ordnance Survey map, facing page 256, the contour interval is 50 ft. (The contour lines are brown.)

Where on the map is this signpost?

The church with a spire, Aldwincle.
On the right is the village school

The church with a tower,
Aldwincle

Using the conventional signs

This is how one boy finds the conventional signs useful. He is going by train to the main station at Thrapston. Then he has to find his way to Aldwincle Post Office, two miles to the north. On the train he looks at his One-Inch Ordnance Survey map and works out his route.

"When I come out of the station I shall turn left down the High Street (A604), and then left again where the church with a spire stands behind the buildings on the corner.

"Leaving the town by this road (A605), I shall soon see, on the left, a reservoir and a river. Then I shall see a railway which goes along an embankment and into a cutting. Beyond the cutting is a copse. I shall walk under a line of pylons which crosses the road, and almost half a mile farther on, just over the top of a small hill, I shall turn left, down a track leading over the railway. The track crosses a river, then a stream, and then some meadows.

"Away to the right I shall see a square-towered church on the edge of Aldwincle.

256

I shall cross a lane and take a footpath leading to the Inn. When I meet the road I shall turn half left, and go past the church with a spire. The Post Office is beyond it, on the right.

"The distance on the map from the station to the Post Office is about three inches, so I shall have to walk about three miles."

When you have followed this walk on the One-Inch map opposite, follow the last part of it on the 1:10 000 map on page 254.

Aldwincle Post Office

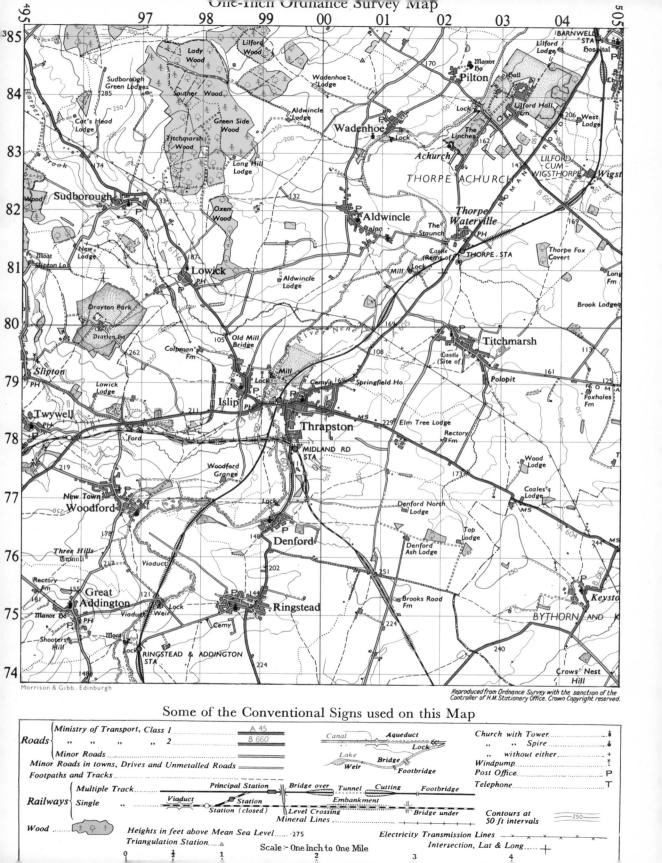

Some of the Conventional Signs used on this Map

Roads
Ministry of Transport, Class 1 .... A 45
" " " 2 .... B 660
Minor Roads ....
Minor Roads in towns, Drives and Unmetalled Roads ============
Footpaths and Tracks ....

Canal   Aqueduct
Lock
Lake
Weir   Bridge
Footbridge

Church with Tower .... ⛪
" " Spire .... ⛪
" without either .... +
Windpump .... 𝕏
Post Office .... P
Telephone .... T

Railways
Multiple Track ....
Single " ....
Viaduct   Station
Station (closed)
Principal Station
Bridge over   Tunnel   Cutting   Footbridge
Level Crossing   Embankment
Mineral Lines ....   Bridge under

Wood ....

Heights in feet above Mean Sea Level ......·275
Triangulation Station .... △

Contours at
50 ft intervals 250

Electricity Transmission Lines ....
Intersection, Lat & Long .... +

Scale :- One Inch to One Mile

Morrison & Gibb., Edinburgh

Reproduced from Ordnance Survey with the sanction of the
Controller of H.M. Stationery Office. Crown Copyright reserved.

The compass

How can you find out where north lies? At midday you can find north by standing with your back to the sun. Then you are facing north. At night, look for the bright North Star, with the Plough pointing to it. You can also find north by looking for the mossy side of trees. If you have a compass you can use it to find which way is north.

A pocket compass. The needle swings until it points to the north

The swinging needle of the compass always points to magnetic north, which is very near to true north. If you *orient* your compass (put the north of the compass card under the north-pointing needle) then you can "read" your map properly.

A small pocket compass does not cost much, but you can make a simple compass by rubbing an ordinary needle with a magnet, pushing the needle through a match stick, and floating it in a bowl of water. The needle will turn and point to the north.

Needle — Magnet

Making a compass by stroking a needle with a magnet

When the needle is floating, it turns and points to the north

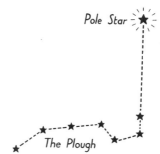

The stars at the end of the constellation called the "Plough" (or "Great Bear") point towards the Pole Star. This star is always due north. (Remember that the Plough appears to revolve round the Pole Star, so it will not always look as it does here.)

Every good map has on it an arrowhead something like this:

When you use the map, you must first point the arrow-head towards the north. Then you can "read" the map correctly, looking to your right for places in the east, to your left for places in the west, and so on.

Sailors "box" the compass by saying the names of the 32 main directions in which a compass points. Here are the first five points: north, north by east, north-north-east, north-east by north, north-east. Can you work out the rest?

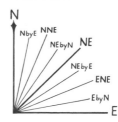

Nine of the thirty-two "points" of the compass

How to describe a position on a globe or map

If you want to point out the position of a desk in a classroom you can do so by referring to the door and the blackboard: "The second desk from the blackboard in the third row from the door."

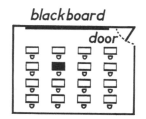

Lines of latitude

In a similar way we can describe positions on a globe. Our first "starting line" is the equator, which goes round the earth. Other lines are imagined to the north and south of the equator. These lines are called lines of *latitude*. They are always the same distance apart. The lines are numbered in "degrees" from 0° (nought degrees) at the equator to 90° at the north and south poles. Thus part of Cornwall is 50° north (50° N.) of the equator. The tip of South Africa is 35° south (35° S.) of the equator.

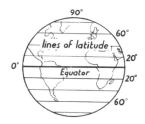

Lines of longitude

Our second "starting line" is a line drawn from the · north pole to the south pole, through London. Other north-south lines, called lines of *longitude*, are on either side

of it. They are widest apart at the equator, but become closer together to the north and south of the equator, until they all meet at the two poles.

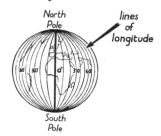

The line of longitude through London is 0°. The other lines of longitude are numbered from 0° to 180°, going eastwards and westwards. On the opposite side of the earth from London is the line of longitude of 180°. The islands of Fiji are on this line.

Halfway between longitude 0° and 180° are the lines of longitude of 90° E. and 90° W. The city of New Orleans, in the U.S.A., is on the line of 90° W. The line of 90° E. passes through the delta of the River Ganges.

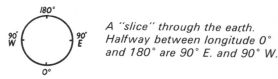

A "slice" through the earth. Halfway between longitude 0° and 180° are 90° E. and 90° W.

By using these lines of latitude and longitude we can refer to the position of places on a globe (or on a map, which is really part of a globe).

Here are the descriptions of two places. Look for them in an atlas:

New Orleans (U.S.A.): 30° N., 90°W.
The mouth of the River Amazon (South America): 0° N., 50° W.

On most of the maps in this book there are lines showing latitude and longitude.

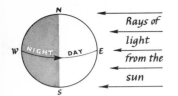

The earth is turning in a west-to-east direction

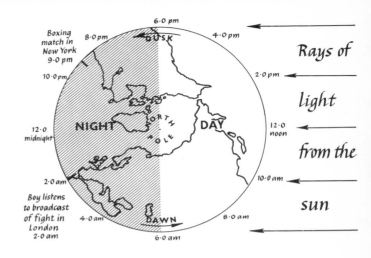

Time

Have you ever listened to a commentary on a cricket Test Match from Australia? If so, you know that when the cricketers are just finishing the day's play, you are having your breakfast. When it is 6.0 p.m. in Sydney, it is 8.0 a.m. in Britain. Similarly a broadcast of a boxing match at 9.0 p.m. in New York is heard at 2.0 a.m. in London.

Why is this? The earth is turning in a west-to-east direction. This means that daylight comes earlier to places in the east. Dawn in India comes before dawn in Africa, but each has its dawn at the same time by the clock.

To achieve this, the world is divided into 24 *zones*, each of which has its own time. Standard time, which is used as a basis for working out time all over the world, is decided in Britain. It is called Greenwich Mean Time (G.M.T.). Places east of Greenwich are ahead in time (eAST=

fAST). Places west of Greenwich are slow in time (West=sloW).

There are 24 time zones, as you can see from the map. If you travel east, you must put your watch *on* an hour every time you cross a zone line, otherwise the sun would seem to rise earlier every day. If you travel west, you must put your watch *back* an hour every time you cross a zone line.

What happens if one plane travels *east* from London, putting its clock *on* an hour at each zone line, and one travels *west*, putting its clock back? When they meet at 180°, one pilot says the time is nearly midnight on Friday, the other says it is nearly midnight on Thursday.

The International Date Line

To sort this out, there is an International Date Line, an imaginary line drawn from north to south through the Pacific Ocean. Travellers going *eastwards* across the line *subtract* 24 hours. Thus the first plane would have two Fridays. Travellers going westwards add 24 hours. The second plane would have no Friday. An easy way to remember this is to say "going west, a day 'goes west'."

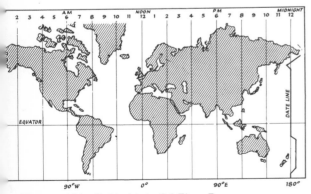

The world is divided into 24 Time Zones

259

THE HOT LANDS
3 Brazil

Brazil is a huge country, the fifth largest in the world. But although it is twenty-seven times larger than the British Isles, its population is only half as big again as Britain's. (Brazil 87 million; Britain 55 million.) Much of Brazil is covered by dense tropical jungle, often called the Amazon forest, because the great River Amazon flows through it. (See the map on page 330.)

The River Amazon

The Amazon flows nearly 4000 miles (6500 km) across South America, from the Andes of Peru in the west, to the Atlantic Ocean in the east. The hundreds of tributaries which flow into the main river, and the land between them, make up the *basin* of the Amazon.

All kinds of craft travel along the waters of the Amazon and its tributaries. Canoes hollowed out of tree-trunks are used by most of the river dwellers, but doctors, and other people who need to travel quickly, usually go by hydroplane.

A hydroplane

Steamers can go a long way up the Amazon, and ocean liners can travel 1000 miles (1600 km) up the river, as far as Manaos. At Iquitos in Peru, 1000 miles further on, the Amazon is still more than a mile wide, and deep enough for cargo boats to dock.

The Amazon forest

The forests steam with moist heat, and swarms of mosquitoes and other insects make life very difficult for Europeans. At one time, few people explored the Amazon basin, but to-day, oil and minerals have been discovered, and oil drillers and engineers are working there.

The Amazon forest

The city of Manaos, 1000 miles (1600 km) up the River Amazon. Fifty years ago, when the world's rubber came from the Amazon forest, Manaos was a "boom city"

The port at Manaos, which can be reached by ocean-going vessels. Floating quays are needed, as the river level changes by as much as 30 ft (9 m), through variations in rainfall over the basin of the river

The hot, wet forests contain many valuable hardwoods, including mahogany and rosewood, both used for making furniture.

Traders go into the forest to collect castor beans, brazil nuts and rubber. The tree which produces rubber originally grew only in the Amazon forest, but nowadays there are rubber plantations in Indonesia, Ceylon and Malaysia. The countries of the Amazon basin, instead of exporting rubber, now produce only enough for their own needs.

Brazil nuts are enclosed in a hard outer covering

The campos and the high lands

To the south of the tropical forests are the *campos*, grasslands similar to the savannah in Africa. Here, very little rain falls, and then only at the hottest time of the year, so that only coarse grass, thorny bushes and

trees can survive. On the better land, the few people who live there rear cattle for meat.

Most of the people of Brazil live on the coastal plain which lies between the Brazilian Highlands and the sea. On this plain, all the crops of tropical countries can be grown: sugar cane, cocoa, cotton, tobacco and rice, as well as fruits such as oranges, limes, bananas and pineapples.

261

A coffee plant showing the fruit, the beans, and the white, starry flowers

Picking the ripe coffee "cherries". A good picker can gather as much as 200 lb (90 kg) of fruit in a day

Winnowing coffee to remove the dust and leaves from the fruit

The coffee beans are spread out in the sun to dry

Coffee

In the south-east of Brazil, the land rises sharply from the coastal plain. More than half the world's coffee is grown on the hillsides, much of it on large plantations (called *fazendas*) west of Rio de Janeiro and round São Paulo. The hillsides there have a rich, red soil.

There is plenty of sunshine, and plenty of rain which drains away down the slopes.

The seedlings are planted out during the summer months, from November to February, in the wettest part of the year. They grow in the shelter of other trees until they are large enough to be planted out again.

From its fifth year onwards the coffee plant produces small fruits which look like crimson cherries. Inside, each fruit has two greyish-green *beans* which are surrounded by a soft pulp. After the pulp has been loosened by squeezing the fruit in machines, the beans are washed in cement troughs, and spread on concrete drying grounds to dry in the sun. (On some *fazendas*, the drying is done indoors by artificial heat.)

When the beans have been sorted, they are put into jute bags and sent to São Paulo. From São Paulo, the "coffee railway" takes them down the steep slope of the Brazilian Highlands to the port of Santos. Most of Brazil's coffee crop goes to the U.S.A.

Rio de Janeiro from the air. On the right is Sugar Loaf mountain

Minerals, factories and a new capital

Brazil is not only a farming country. In the Brazilian Highlands, there are rich deposits of manganese, gold, diamonds, mica and iron-ore. Though Brazil has little good coal to give power she has many swift streams which are used to make electricity for industry.

Brazil has changed rapidly in the last twenty years. Hundreds of new factories have been built to manufacture everything from aeroplanes to jewellery, and from diesel trains to shoes. Towns such as São Paulo and Rio de Janeiro have grown into bustling, modern cities, with skyscrapers and airports.

Every year, more people from the country areas have been moving to the towns, or to the fertile coastal plain where it is easier to earn a living. As Rio is too hemmed in by hills to be able to expand, a new capital city, Brasilia, has been built on a lonely plateau 600 miles (nearly 1000 km) north-west of Rio.

Ultra-modern architecture in Brasilia—a chapel

263

4 Pakistan and Northern India

For many years Great Britain governed India, but in 1947 British rule came to an end and the separate republics of Pakistan and India were formed. Pakistan is a Muslim country; in India most of the people are Hindus, though there are many Muslims, Christians and Sikhs. Both republics are members of the Commonwealth.

Pakistan consists of two separate parts: *West Pakistan* includes the River Indus, the seaport and airport of Karachi and Islamabad, the new capital. *East Pakistan* is 1100 miles (1700 km) away on the eastern borders of India.

In Pakistan and Northern India there are many different kinds of scenery. In the north are the mountain ranges of the Himalayas. Because they are so high, they are covered with snow and ice. Farther south, there are broad plains made by the Indus, Ganges and Brahmaputra rivers.

In West Pakistan, the plains made by the River Indus and its tributaries are called

A Muslim from the North West Frontier of Pakistan

A Hindu—notice his caste mark

West and East Pakistan and northern India. Delhi is the capital of India; Islamabad is the capital of Pakistan

the *Punjab*. The West Punjab belongs to Pakistan, and the East Punjab to India.

The climate
Instead of spring, summer, autumn and winter, there are *three* seasons.

1 *The hot, dry season: from March to May* In the hot season, temperatures in many places rise to over 110° F (43° C) for some hours during the day. There is no rain, and the intense heat burns up the earth.

2 *The hot, rainy monsoon season: from June to September* Towards the end of June, south-west winds blowing across the Indian Ocean bring heavy rain. These winds, which blow at the same time each year, are called *monsoon* winds. At the beginning of the monsoon there are severe storms, with thunder and lightning. After the monsoon has "broken", it rains every few days, and the hot, steamy air makes everything feel damp. There is so much moisture in the air (*humidity*), that matches will not strike.

A shepherd and his sheep in the Punjab

3 *The cool, dry season: from October to February* In October, the winds change and blow from the north-east, from land to sea. Although this is the *cool* season, the winter temperature of some places in India and Pakistan is higher than the summer temperature of Great Britain.

The dry lands

This is what a Pakistani boy wrote in his diary about his home in the West Punjab.

> May 30th. The midday temperature for the last two days was 110° F (43° C). Today it is nearly 115° F (46° C). The heat this year is so bad that there are cracks in the ground a foot wide, the animals are dying, and the countryside is brown and scorched.

The Punjab is so far inland that the monsoon may reach the dusty plains for only two months each summer. In a year of "bad" rains there is not enough food to go round, and during droughts many families are hungry.

Today new dams have been built, so that the farmers in the dry lands do not have to depend solely on the monsoon rains to water their crops.

Even with irrigation, much of the land is too dry for rice-growing; instead, wheat is the main winter crop. It is sown after the south-west monsoon, and harvested in March, before the great heat.

Millet, which needs very little rain, is grown as a fodder crop for animals, as well as for food. Cereals, together with vegetables, are the chief food of the people of India and Pakistan. Wheat and millet are often made into a kind of porridge by grinding the grain between two flat stones, and then cooking it in water over a fire.

Thousands of Hindu pilgrims come to wash away their sins in the Ganges. Families bring cooking pots and camp for days on the main Ghat. Many bring soap and have a bath too

Cotton is the most important summer crop. Some of the cotton is spun and woven in the homes of the villagers; some is sent to modern mills. Raw cotton grown in India goes mainly to mills in Bombay where it is manufactured into cotton cloth. Raw cotton grown in West Pakistan goes mainly to Karachi where it is exported. There are no mills in Karachi, for there the air is very dry, and if cotton is spun in too dry a climate, the threads break.

The River Ganges

The Ganges rises in the snow-fields of the Himalayas. Hundreds of streams rush down the mountains to join the main river, carrying large quantities of mud and gravel with them as they flow along. Every year, during the monsoon rains, the Ganges and its tributaries overflow their banks. The water floods across the plains, spreading a new layer of mud on the dry earth.

When the water has gone, crops of rice, sugar cane, red peppers and other spices grow well in the soft, warm mud.

The Ganges, and its tributary the Jumna, are holy rivers to the Hindus. On their banks are some of the oldest cities in India —Delhi, Agra, Allahabad and Varanasi (formerly called Benares). At Varanasi, the steps (ghats) lining the river are often crowded with Hindu pilgrims who have come to bathe in the sacred Ganges. Sick people, too, are brought to the river to be healed, and the ashes of the dead are thrown into its current.

The changing countryside: farming

Three-quarters of the people work on the land, living as they have always done, in tiny villages amidst their fields. In the past, most of these people were poor farmers.

Many houses are built of sun-dried mud bricks. Notice the string bed, the sacred cow, and the new pump

Today, the old India is changing, and throughout the country the people are becoming better farmers. "Village helpers" are sent by the government to show the people how to use artificial fertilisers, and how to sow new varieties of seeds, so that they can grow more food on the same land.

Here are some ways in which the villages are being improved:

Making drains and paving the streets.
Filling in mosquito breeding holes.
Repairing and building new wells to make a "safe" water supply. (Typhoid germs, which cause a dangerous fever, can be taken into the body in water which is not pure.)
Building schools and health centres.

If the people of the villages are to have better food, animals as well as land must be made to produce more food. In Great Britain, where there is enough rain to provide plenty of good grass, the cows give eighteen times more milk than those in India. (An Indian cow is an ox-like animal with a large hump.) India has far more cattle than the land can feed, but because cows are generally thought to be sacred, the people do not kill them.

Indian cow 40 gals (180 l) a year   *Buffalo 100 gals (450 l) a year*   *British cow 720 gals (3300 l) a year*

Nowadays, some farmers are crossing their cattle with imported breeds so that they will give a greater quantity of milk. Every year more milk is needed to feed the increasing number of people in the cities.

The country of the Ganges delta

Towards the sea, the Ganges divides into many mouths. The mud carried in the water of the river sinks to the bottom and in time a flat piece of land is built up. (Land such as this, which forms between the mouths of a river, is called a *delta*.) Pakistan and India share the delta of the Ganges.

As there is a heavy rainfall round the Ganges delta, and plenty of water, the vegetation is always green, even in the hot season. Formerly much of the delta was covered with marshy jungle inhabited by tigers and crocodiles. Nowadays, parts of the delta are used for growing rice. The country is a maze of waterways: the houses are on earth mounds, and the few roads and railways are built on high embankments.

Jute growing

Nearly all the world's jute is grown in the hot, swampy delta country of Pakistan and India. Jute is sown in February or March and harvested at the end of June, when the plant has grown to about fifteen feet high. The stalks are cut, bound in bundles, and left under water in streams and ponds.

As soon as the woody, outside stem begins to soften, the long tresses of fibre inside are stripped off and cleaned by beating them on the water. The people stand waist deep as they work. Then the long hanks are hung on racks to dry.

Cutting and soaking jute

Stripping the fibre and hanging it in the sun to dry

Jute manufacture

Indian jute is rolled into bales and taken to Calcutta, sometimes by bullock carts, but more often in flat-bottomed boats called "jute flats". The jute mills lie on both banks of the River Hooghly, and stretch for many miles up-stream from Calcutta. In the mills, the raw jute is spun into yarn; most of the yarn is woven into different kinds of sacking.

A new building in Pakistan. The shuttering round the windows throws shadows, and so helps to keep the building cool

potter making earthen cups on a wheel

Calcutta—the Howrah railway station

The jute grown in Pakistan goes to mills near the Delta town of Dacca, or to the port of Chittagong. East Pakistan sells raw jute to India, which does not grow enough jute for her own factories. Some jute goes to Dundee in Scotland.

As well as being used for sacking, jute fibre is made into garden twine, window sash cords, brown wrapping paper, backing for linoleum and chair webbing.

Calcutta

Calcutta, the capital of the state of West Bengal, is a busy manufacturing city on the River Hooghly, 80 miles (130 km) from the sea. Calcutta's chief exports are tea (grown on the hillsides of Assam near Darjeeling), coal, and articles made from jute.

Calcutta, like many other cities in India, Pakistan and other parts of the world, is very overcrowded. There are many poor people who have no homes, and who have to sleep on the pavements.

To make the city less crowded, a new town is being built with its own industries. It is difficult to use the land around Calcutta because of the marshy country, but there are plans for pumping out the water from two salt lake swamps on the east of the city. Once the swamps have been drained, the land will be used partly for building and partly to grow rice.

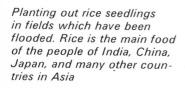
Planting out rice seedlings in fields which have been flooded. Rice is the main food of the people of India, China, Japan, and many other countries in Asia

Building a new steel works, in West Bengal

The changing countryside: industry

The population of India is growing so quickly that agriculture cannot provide enough work for everyone. Other work, besides farming, must be found, and so more and more factories are being built.

Nearly every industry needs steel, and in countries with growing industries a great deal is needed. To the west of Calcutta, in the provinces of Bihar and West Bengal, there are large quantities of good quality iron ore as well as plenty of coal to make it into steel. This part of the country is an important industrial area, with factories making artificial fertilisers, brass and aluminium goods, and cement for the new towns and blocks of flats.

India already has steel works in Bihar and West Bengal, and she is using her steel as fast as she can make it. Three new plants have been built: by Russia, Germany and Great Britain. The steel works built by Britain is in West Bengal, on barren wasteland where the people have a hard struggle to grow crops.

The blast furnaces were made on Teesside, and then shipped in sections to Calcutta.

Every new steelworks and every new factory built in India means regular work and better wages for the people in the nearby villages.

India in the future

India is still largely a country of peasant farmers, but gradually, and slowly, more people are becoming factory workers and traders. India needs more factories, and more goods to sell abroad, so that she can earn more money. Then she will be able to build the new schools, hospitals, houses, roads and railways which are so badly needed.

Whenever a new factory is built, much of the cement and other materials needed are carried by women, in dishes balanced on their heads

5 Australia

Advice to immigrants

"As a passenger, you will have to provide your own bedding, and also whatever candles you burn in your own cabin. If you are a man given to reading, I would advise you to buy a large transparent lantern, six pounds of wax candles and some amusing books."

This was the advice given to some of the first British immigrants to Australia, over 100 years ago, when they were preparing for the long sea voyage.

How different it is for emigrants to Australia today. The journey by sea from Britain to Adelaide takes 26 days instead of four months. The special fare for immigrants is £10 for each person, and children travel free.

Size and population

Australia is 32 times the size of Great Britain, and yet in Greater London there are almost as many people as in the whole of Australia. That is why immigrants are needed to develop Australia.

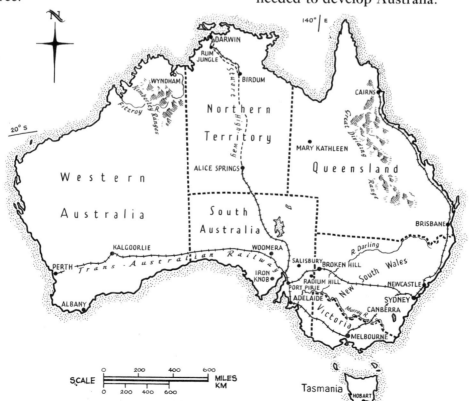

271

*A champion
merino ram*

Much of central Australia is desert. The main cities are in the south-eastern coastal regions of South Australia, Victoria, New South Wales and Queensland, and in the south-west corner of Western Australia. The south-east trade winds bring rain, so the narrow coastal belt, which lies mainly between the mountains of the Great Dividing Range and the sea, is an important part of Australia. Dairy cattle are pastured there, and sugar-cane, tobacco and tropical fruits are grown in the hotter climate of Queensland.

On the plains of New South Wales and Queensland many sheep are grazed. Other sheep-raising areas are Victoria, South Australia, and the south-west of Western Australia. Some sheep such as *merinos*, with their soft, crinkly wool, are reared for their fleeces. Other breeds are reared for their good quality meat.

The government does not want all the immigrants to go to the eastern states, or to the cities. Already more than half the people in Australia live in the state capitals such as Sydney, Melbourne and Adelaide. Immigrants are also needed in the smaller towns, and in the "outback", the vast, lonely stretches of country away from the towns.

Aeroplanes in the outback

The roads over much of the north are merely rough tracks, passable only during the dry season. There is one railway line from Perth to Port Pirie, west to east across the Nullarbor Plain. A railway links Port Pirie with Alice Springs and an all-weather road runs north from Alice Springs to link up with a railway to Darwin.

In the outback, as a result of this shortage of road and rail links, aeroplanes are used a great deal. Many farmers and doctors use their own light planes. Flying is easy in a country with such wide open spaces and such good weather, and very useful for men whose nearest neighbour may be a hundred miles away.

Aborigines

On the cattle stations, many *aborigines* work as stock boys, breaking in the wild horses and rounding up the cattle. The aborigines (the earliest known inhabitants of the country) arrived in Australia between ten and twenty thousand years ago. Nobody really knows where they came from, but it is thought that they may have travelled from southern India.

Even today some of the *myalls* (bush aborigines) live as they have done for thousands of years. They sleep in rough shelters called *gunyahs*, they never cultivate plants for food, and their only animals are half-wild dingo dogs.

The myalls are really "stone-age" men. They rub two sticks together to make fire, and hunt for food with a spear, woomera and boomerang.

Many of the station aborigines have been educated in mission or government schools, and in some parts of Australia attempts are being made to educate the myall children as well, and particularly to teach them English. In the Northern Territory, teachers with jeep-drawn caravans travel round with the nomadic tribes, setting up schools at the water-holes or in the shade of a tree.

Two aborigines with their weapons. The man on the left has a boomerang, the one on the right a spear and woomera

An aborigine car mechanic

Shell
to weight end of woomera

four-pronged spear

The rocket range town of Woomera is named after the spear-thrower used by aborigines

Flying over the "Dead Heart"

Mr. Webster, an English engineer, has emigrated to Australia to work on rocket research at Woomera. With his family he arrives by boat at Adelaide, and then flies north to Woomera.

The first part of his journey is over dry country which has barely enough grass for sheep to be pastured. Soon the plane crosses the deserts of South Australia. The desert lands of Australia stretch from the far west of New South Wales and south-west Queensland, into part of South Australia, Western Australia and the Northern Territory. Everywhere is so dry and barren that this part is often called the "Dead Heart". For hundreds of miles the flat, dusty plains are covered with foot-high salt bush and clumps of thorny-leaved spinifex grass.

Woomera

The town of Woomera has been built in the desert of the "Dead Heart". The rocket range, for testing guided missiles, stretches for over 1000 miles. This part of Australia was chosen for the range because no people live there, and because there is little possibility of future settlement, owing to the shortage of water.

The baobab, or bottle tree, can live in lands with little rain. It stores up to 80 gallons (360 l) of water in its trunk

Alice Springs

Scrub country in the outback, near Alice Springs. The earth is red and sandy; the trees are the hardy mulga; the ground is covered with salt brush

Many problems were solved before people could make their homes in Woomera.

Water was brought over 100 miles by pipeline, and thousands of young trees were planted to provide shade, and to prevent the top soil from blowing away. An airport was made, and a railway line built to join the Trans-Australian line from Perth to Port Pirie. This railway is needed for the transport of stores, particularly perishable foods such as milk, butter and meat.

Mrs. Webster finds that most of the food in the shops is grown in Australia. Australia covers such a large area, that there are several different kinds of climate. Both tropical and temperate crops are grown.

In the shops Mrs. Webster can buy sugar and tropical fruits which were grown in Queensland, apples and pears from the cooler island of Tasmania, and grapes, wine and dried fruit from the vineyards of South Australia. Her family eat Australian-produced lamb, pork and beef, and wear clothing made from the wool of the merino sheep. Even the furniture in her new home is made from Australian wood—from the eucalyptus (gum) tree, and from woods such as walnut, maple and cedar which grow in the coastal jungles of Queensland.

Rockhampton, Queensland—the centre of a mining and industrial region

275

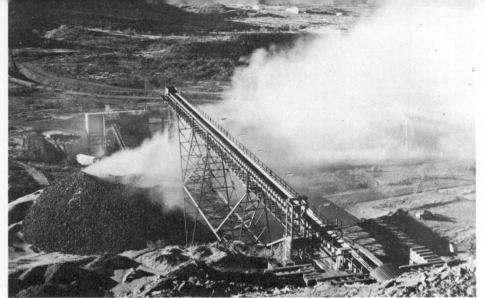

An iron ore mine in Western Australia   Surf riding

A miner in Newcastle

Mr. Hughes, a miner from the Rhondda Valley, has been in Australia for several years. He is now working in Newcastle, a port and industrial town 100 miles (160 km) north of Sydney.

Newcastle, which is on the New South Wales coalfield, produces good quality coking coal. Some is used in the city's blast furnaces, and some is sent by sea to the gasworks in the industrial section of Sydney Harbour, and to Port Pirie (South Australia) for the lead smelting works.

There are many other places in Australia where Mr. Hughes could work as a miner. Gold is mined at Kalgoorlie (Western Australia), iron at Iron Knob (South Australia), and silver, lead and zinc at Broken Hill, New South Wales. There are rich deposits of uranium at Radium Hill in South Australia, and at Mary Kathleen in Queensland.

In summer, when Mr. Hughes has finished his shift in the mine, he often goes surfing at one of Newcastle's beaches. Sydney's Bondi and Manly beaches are the best known, but there are hundreds of miles of magnificent beaches stretching along the Pacific coast, from New South Wales into Queensland.

Sunshine and sport

Australia has a great deal of sunshine. About a third of the continent, north of Alice Springs, is in the tropics. Perth, the sunniest state capital, has an average of almost eight hours' sun a day.

In Australia, thousands of city dwellers live within a short bus or train ride of the beaches. In the summer, at weekends and at holiday times, the people flock to the sea, or go fishing, bush walking, horse-racing or sailing.

Although the country has a population of only 12 million, the Australian's love of sport, and the long sunny summers, have produced many of the world's leading cricketers, swimmers, athletes, golfers and tennis players.

6 The Southern States of the U.S.A.

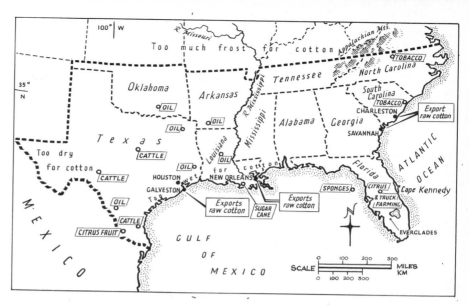

The United States of America is a huge country of forests and mountains, deserts, plateaus and vast plains. Most of the country has a temperate climate (*temperate* means moderate or mild). You will read about the temperate parts of the country in Chapter 14.* In this chapter you will read about the Southern States, which have a hot climate.

Here is a map of the Southern States. In the east, there is a range of mountains called the Appalachians, but most of the states consist of plains. The River Mississippi,

* On page 318 is a map of the U.S.A.

The mechanical cotton picker works like a vacuum cleaner

which is over 2000 miles (3200 km) long, flows through the centre of these states into the Gulf of Mexico.

Fertile land, a long growing season, and warm, sunny weather all help to make agriculture the chief occupation of the people of the south.

Crops

Cotton is the most important crop. It grows well from eastern Texas to North Carolina, where there is rarely any frost. The ripe bolls are picked by machine or by hand, and taken to a factory where the seeds are torn from the fluffy lint. Three-quarters of the cotton crop is used in the country's textile factories; the remainder is exported. The chief ports for the export of cotton are Galveston and New Orleans on the Gulf of Mexico, and Charleston and Savannah on the Atlantic cost.

Soya beans, *groundnuts* (peanuts) and *maize* are also grown by the farmers of the south. They sell the maize to poultry farmers; the oil from the groundnuts and soya beans is used to make margarine.

277

A truck farmer and his son, from Louisiana, with a load of shallots (green onions)

The tobacco farmer removes the flower bud and top leaves to strengthen the rest of the plant

Tobacco is grown mainly in the south-eastern states of North and South Carolina, Georgia and Florida. These states have factories where different varieties of tobacco leaves are made into cigarettes, cigars and pipe tobacco.

Along the low-lying coast of the Gulf of Mexico, the land is too wet and marshy for cotton, but the moist climate and hot summers (over 80° F—27° C) provide the right conditions for growing *rice* and *sugar cane*. Most of the sugar cane is grown near New Orleans, in the delta of the River Mississippi.

Citrus fruit
Citrus fruits, especially oranges and grape-fruit, are grown in Texas and in Florida. These fruits are sent all over the United States, as well as to Great Britain and other countries whose climates are too cold for "sub-tropical" crops. Some of the fruit is sent to juice-canning factories. The pulp and rind are dried and made into fertilizers and cattle food.

Truck farming (market gardening)
Truck farmers in the Southern States, such as Florida, where the winters are mild, grow early vegetables for the industrial towns of the north-east. (Truck farmers take their name from the trucks—lorries—which carry their produce to the towns.) The main crops are potatoes, tomatoes, green vegetables and onions.

Cattle rearing
The climate becomes drier towards the west. Western Texas, for instance, is too dry to grow cotton. Instead, vast herds of cattle are pastured on miles of rolling grasslands. The biggest cattle ranch in North America is in Texas; it covers over a million acres, a greater area than the English county of Hampshire.

A musk rat   *A skunk*

A sponge diver at Tarpon Springs, Florida

Trapping

In places along the Gulf Coast, as well as in parts of Florida, there are swamps where trappers hunt musk rats for their reddish-brown fur (musquash) as well as skunks, otters, mink and other fur-bearing animals.

The largest swamp, the Everglades, is in the south of Florida. Here, besides alligators and poisonous water snakes, there are thousands of beautiful birds: pelicans, fish-eating hawks, storks and large herons.

Forest industries

On sandy land in the south there are forests of pines. Some of the pine wood is pulped to be used in the manufacture of paper,

rayon and plastics; some is heated in closed furnaces to extract the tar which it contains. The pines also have a sticky, yellowish sap from which turpentine is obtained.

The holiday "industry"

Because of its sunny climate, Florida has many popular holiday resorts. One of the biggest is Miami, where the holidaymakers stay at fine hotels and lounge on the golden sands.

Cape Kennedy, Florida—an Apollo spacecraft sets off to the moon

Miami, Florida, is a popular holiday resort for Americans

Oil

In Texas, Oklahoma, Louisiana and Arkansas there are rich oilfields. The oil is pumped from the rocks underground. Some of it is refined in Texas and Louisiana, but because it is more difficult to transport petrol than crude oil, most of the crude oil is pumped through pipes to the ports of Galveston and Houston on the Gulf coast, and taken by tankers to refineries near to the places where the oil will be used. Much of the oil goes to the Atlantic ports of New York, Baltimore and Philadelphia where there are refineries serving the densely populated north-eastern states. Some of the oil is taken to Britain.

Industries

At one time, all the raw cotton from the Southern States was sent to the industrial

Coupling the hose through which Texas oil will be pumped into a tanker

cities of the north-east to be manufactured into cotton cloth. Today more than three-quarters of the country's cotton mills are in the south. Most of them are in North and South Carolina and in Georgia, where hydro-electric power is generated by the streams of the Appalachian mountains.

Many of the other industries of the south have developed because the raw materials are close at hand. There are sugar refineries, rice husking mills and fertilizer plants. Rayon is made from wood pulp and cotton waste, and insulating material from sugar cane waste.

Although more and more factories are being built in the south, it certainly cannot be called an "industrial area". The Southern States produce such huge quantities of raw materials that even after supplying manufacturers in other parts of the U.S.A., they still have plenty to export.

The waterfront at New Orleans, Louisiana, near the mouth of the Mississippi River

7 The Fiji Islands

In the last hundred years, the world has changed rapidly. Never before have so many countries made such rapid progress. In a comparatively few years Russia, Japan, Australia, the U.S.A., and other countries have changed from undeveloped lands with primitive farms into bustling modern countries with efficient farms, modern factories, busy ports, and huge cities.

Other countries have changed little, and the lives of the people are much as they were a hundred years ago. This is so on the tropical islands of Fiji in the Pacific Ocean. These islands are peopled partly by the Fijians, a dark-skinned, fuzzy-haired people, and partly by the descendants of Indians who went to Fiji to work on the plantations. Today many ships travelling between New Zealand and the Panama Canal call at Suva, the capital of Fiji. Trans-Pacific planes call at the international airport of Nadi.

The Islands of Fiji

Although there are over three hundred islands, only about a hundred are large enough to be inhabited. There are three main types of islands.

A Fijian policeman wearing a cotton kilt called a "sulu"

1 *The High Islands* were formed long ago by volcanoes erupting. They have a fertile soil, wide rivers, and mountains covered with tropical rain forests. The soil of volcanic islands is usually rich in minerals; on Viti Levu, the largest island, gold and manganese are mined.

2 *The Low Islands* are made of coral* which has been ground into white sand by the beating of the waves.

3 *The Coral Atolls.* Islands made of a circular ridge of coral are known as atolls.

* Coral is a hard material made from the lime in sea water by colonies of sea creatures called polyps.

A coral atoll

A rough sketch map of the atoll

A village on the island of Viti Levu

Coral atolls are only a few feet above sea level; very little will grow on them. The people who make their homes on atolls live mainly on coconuts, fish and the eggs of sea birds. Rain-water is collected in "tanks" sunk in the sand, or in the hollows scooped in the forks of trees. The calm water enclosed by an atoll is called a lagoon.

Life in a Fijian village

Most Fijians live in villages away from the towns. As there is a rainy season (from December to about the end of March), the houses are built above the ground on blocks of stone. They are thatched with dry coconut leaves or sugar-cane leaves, and the floor is covered with woven mats.

The Fijians who live by the sea eat a great deal of fish. They share a communal net or use spears with iron points. When they travel through the reef passage to the small, uninhabited islands to collect coconuts or turtles' eggs, they go in a canoe made from a hollowed-out tree-trunk. The sail is made of matting, and to keep the boat steady, a framework of lighter wood, called an *outrigger*, is built out from the side of the canoe.

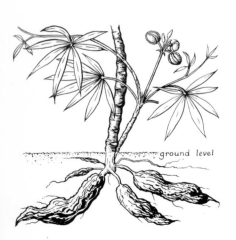

Left: *Cassava, also known as manioc, is a shrub with thick tubers. These tubers are washed, sliced and boiled before being eaten. Tapioca is made from cassava*

Right: *Breadfruit is the large fruit of a tree which grows 50 ft (15 m) high. It is cut unripe and baked. When the rind is removed the inside is rather like bread*

Coconut growing

Many Fijians are farmers, working on coconut plantations, or growing coconuts on their own land. When the nuts are cut from the trees, they are enclosed in a brown husk two or three inches thick. The islanders split open these husks and use the fibre as fuel for their fires.

The most important part of the nut is the *kernel*. The kernels are dried in the sun, on racks over smouldering fires, or in mechanical driers. When the white coconut flesh shrinks and becomes brown in colour, it is called *copra*.

In the picture you can see a copra collecting boat. The sacks are ready to be loaded into the hold and taken to Suva, where the oil is squeezed from the copra for use in the manufacture of soap and margarine. Both copra and coconut oil are exported to Great Britain.

The sugar cane industry

A great deal of sugar cane is grown on Viti Levu, but very few Fijians earn their living in the sugar industry. Most of the cane growers and workers in the crushing mills are descendants of the Indians who were brought to Fiji 80 years ago. Today, there are more Indians than Fijians in the Fiji Islands.

Banana farming

Some Fijians grow bananas, planting them in fields beside the rivers. The bananas are cut while green, loaded on to bamboo rafts or flat-bottomed punts, and taken down the rivers to packing stations. Nearly all Fiji's bananas are bought by New Zealand.

Removing coconut kernels from their husks

The copra collecting boat calls at the islands two or three times a year

Bamboo rafts are used to transport the bananas

283

A village primary school

A botany class

Education in the Fiji Islands

Most Fijian villages have a primary school for the children of the village. When it is too hot in the classroom, the children have their lessons outside, in the shade of the coconut trees.

The secondary schools are mainly on the island of Viti Levu. Nearly every secondary school is a boarding school with a farm attached where the children grow the food they eat, such as bananas, breadfruit and the starchy roots of the *cassava* plants. Fijian children are good at games. The boys play football without boots, and cricket in cotton kilts called *sulus*. (On page 281 there is a drawing of a Fijian policeman wearing a *sulu*.) The girls play rounders and basketball.

More and more children in Fiji and the other British Pacific Islands are going to the secondary schools, so that afterwards some of them can train to be doctors, nurses and teachers.

The island of Bau, home of a family of chiefs

The Coat of Arms of Ghana

8 Ghana

More than four hundred years ago, European traders visited the Guinea Coast of West Africa. The gold dust they bought from the negro chiefs was so pure that coins (Guineas) made from it were worth 21 shillings instead of 20. For a long time part of this coast was called the Gold Coast, and was ruled by Great Britain. Today, although the country is still part of the British Commonwealth, it is governed by its own people, and the name of Gold Coast has been changed to Ghana. (Ghana is the name of an ancient African kingdom which has long since disappeared.)

Surf boats and harbours

The coast of Ghana is fringed by a belt of

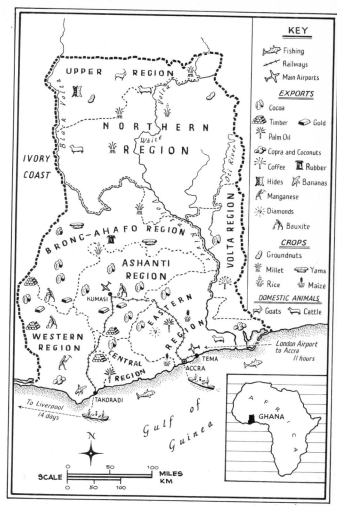

The products of Ghana

Surf boats are still used to take cargoes out to ocean-going ships

surf which breaks heavily on the beach, and there are no inlets large enough to take ocean-going ships. Instead, ships anchor beyond the breakers, and surf boats ferry their cargoes ashore. This even happens at Accra, the capital.

But soon the surf boats may disappear from this part of the Gulf of Guinea. Takoradi and Tema have man-made harbours, each with an artificial breakwater to stop the force of the Atlantic rollers.

285

*A new housing estate at Kumasi in the Ashanti cocoa-growing area.
The ditch at the side of the road carries away the heavy rains of the
rainy season. The trees are pawpaws, which have a juicy, golden
fruit*

In Accra, as in other towns throughout Ghana, tall, concrete
office blocks and self-service stores are being built alongside
the native markets. At Kumasi, the chief town of the Ashanti
cocoa-growing area, and in other towns, there are new housing
estates, schools, and colleges like the one in the above picture.

Cocoa

Cocoa is the most important crop grown in Ghana. Cocoa and
chocolate are made from the beans of the *cacao* tree. This tree
must have shade, heat and the heavy rainfall of tropical
forests. In the south of Ghana, in the Ashanti forest region,
there is ideal country for growing cacao trees.

*Cutting the pods from the
cacao tree. They grow from the
trunk as well as from the main
branches*

Kofi (his name means Friday) is an African farmer who earns
good money by cultivating cacao trees. Most of the cacao crop
is grown on small farms owned and worked by men like Kofi.
He cuts the lower pods from the tree with a sharp cutlass, and
uses a knife on a long pole to cut down the higher pods. The
pods are yellow when ripe, with a leathery rind about half an
inch thick.

286

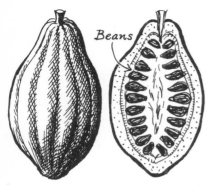

Kofi splits the pods in half, takes out the beans, which are surrounded by a sticky pulp, and covers them with banana leaves to protect them from the rain. Then they are left for a week to ferment, or "sweat". This improves the quality of the cocoa.

The beans are covered with banana leaves

Inside each pod are twenty to thirty cream-coloured beans

Spreading the beans on raised trays, so that they will dry in the sun

Cranes on the dockside load the sacks of beans into the holds of the ship

The beans are taken to the compound and spread on tables where they dry in the hot sun for several days. Then they are poured into sacks and taken by lorry to Tema or Takoradi where they are loaded into the holds of cargo boats and exported.

The diagram on the right shows the year's work of the cocoa farmer. He gathers two crops, a main crop between October and December, and a very much smaller crop between March and May. His work is regulated by the seasons of the year. The dry season is followed by the "big rains", which are followed by the "little rains". Between the harvests, much work has to be done on the farm.

The year's work of a cocoa farmer

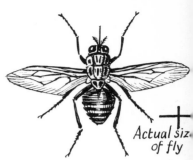

Actual size of fly

Above: *The tsetse fly is related to the house fly. On the front of its head is the "needle" used for skin piercing and blood sucking*

Left: *Setting up irrigation pipes in the savannah*

The savannah

Beyond the forest region, and stretching for hundreds of miles to the southern edge of the Sahara desert, there is open grass country, or *savannah* land. Here, in the hot, dry Northern and Upper Regions of Ghana, cattle are reared and yams, maize, millet and groundnuts are grown.

Although the Black and White Volta rivers never dry up completely, there is usually a great shortage of water, and if the rainy season is late, the smaller rivers and water-holes become dry. In some areas, dams have been built on the tributaries of the Volta to hold back a supply of water for irrigation. Water in the dry season allows the people to grow rice and tobacco.

Palm oil

Palm oil is obtained from the fleshy outer coverings of the fruits of oil palm trees.

Palm kernel oil is obtained from the kernels. Both these oils are used in soap and margarine making.

The tsetse fly

The African farmer's greatest pest is the tsetse fly. This insect, common over all tropical Africa, makes it difficult for him to keep cattle and domestic animals, for its bite causes a disease which makes the cows thin and weak. The tsetse fly also attacks human beings, who then suffer from a disease called "sleeping sickness", which makes them continually drowsy and unable to work, until finally they lose consciousness and die.

Great efforts are being made to wipe out the tsetse fly. The undergrowth, where the females breed, is being cleared, and pools of stagnant water are being sprayed with insecticides, so that the Africans can return to areas where it was once unhealthy to live.

288

A kibbutz at Yotvata in Israel (see page 292)

Sufferers from leprosy being treated at a "Land-Rover" clinic. Some of them may have walked as far as ten miles for treatment

A gold miner drilling a hole so that the ore can be blasted out. Early miners "panned" for gold in river beds. They washed the light sand from the heavy gold

In parts of Ghana, particularly in the north, people still suffer from leprosy. There are some leper settlements, but many patients live normal lives in their villages, and have weekly treatment from "Land-Rover Clinics" which travel through the country districts. With the use of modern drugs, patients are often cured after two years' treatment.

Ghana, like many countries in Africa, is changing rapidly. Many Africans are moving away from their villages to work in the towns where there is electricity instead of paraffin lamps, and piped water instead of a well.

The export of gold, and mahogany logs from the forests, has helped to make Ghana prosperous, but the real wealth of the country is in cocoa. Hundreds of miles of roads and railways, schools, hospitals, three universities—all these could not have been built without the money that cocoa has brought to the country.

A village standpipe. The bowls are made from the fruits of the calabash tree

A farming settlement in the desert beside the Dead Sea. Fresh water for drinking and irrigation is piped from a natural spring

9 Israel

Israel is a small country at the eastern end of the Mediterranean Sea.

Formerly part of the country known as Palestine, Israel has a long history. At various times it has been the "promised land" of Moses, the "Holy Land" of the Christian World, part of the Roman Empire, and part of the Turkish Ottoman Empire. From 1917 it was ruled by the British, until the independent state of Israel was proclaimed in 1948, and Jews all over the world rejoiced that once again they had a national home.

The people

Most of the people of Israel are Jews. Two-thirds of them have arrived since 1948, from Europe, from the Middle East, and from North Africa, to build their homes in a new country. Over 300,000 Arabs still live in Israel. Many of them live in the north, in their own towns and villages.

The country and its climate

The countryside is very varied: there are fertile valleys, such as the Vale of Esdraelon,

ridges of rocky hills in Galilee, Judaea and Samaria, and great stretches of desert in the Negev. The Dead Sea, which lies to the east of the country, is 1300 ft (400 m) below sea level. Seven rivers flow into it, but none flow out: the water is lost by evaporation. This makes the water of the Dead Sea so salt that it is impossible to sink in it.

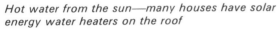
Hot water from the sun—many houses have solar energy water heaters on the roof

Part of the waterline from the Sea of Galilee to the Negev Desert

From April to October the weather is bright and sunny. It is hotter than Britain in summer, particularly when the hot dry wind blows from the desert of Arabia. Though rain and snow fall during the cold season, the winter is not very long, and shortage of water is a great problem for farmers in Israel.

Farming in a difficult land

When new immigrants arrive in Israel, they are taken to their permanent homes and given work in agriculture, in a factory, or in an office.

Many of the new settlers work on the land. Parts of Israel which were once desert are being made into farm land by irrigation and skilful farming. Sometimes, the settlers had to drain swamps, but usually, in this dry land, the problem is to bring water to the desert. Where they succeed, they plant grass and trees to bind the soil, so that winds and sudden rain-storms cannot carry it away.

Rain is heaviest in the north of Israel, and pipelines from the Sea of Galilee in the north and the Yarkon River in the centre of the country carry water to the dry desert of the Negev, in the south. In the coastal plains there are some places which have springs and underground water. As well as irrigating the land, the people terrace the hillsides so that heavy rain does not rush down them, carrying away the soil. They also dam the rushing streams which dry up in all but the wet weather. Scientific methods, heroic hard work and the use of modern machinery are enabling the Israelis to grow crops on land that was desert until a few years ago.

Jaffa oranges and blossoms

The people of Israel now grow three-quarters of all the food they eat. They produce all the vegetables, potatoes, eggs and fruit they need, and are able to export some fruits. Grapefruit, grapes and the famous Jaffa oranges are sent to the ports of Haifa and Ashdod, to be exported to Britain and other countries.

In the hotter, wetter parts of Israel, such as the Jordan Valley, pomegranates, guavas and mangoes are grown. New crops such as sugar beet, cotton, groundnuts and tobacco are also grown successfully, and exported.

Date-palm trees grow in hot, dry deserts wherever there is underground water. Dates are brownish when ripe and are valuable food for men and animals. The dates we eat have been dried in the sun

The kibbutz

There are not many ordinary villages in Israel. Most people who work on the land live in various kinds of cooperative villages. The most common is the *moshav*—in a moshav each family owns its house and plot of land, but buying and selling are done cooperatively, by the village as a whole.

A crate of Jaffa oranges

The best known type of settlement in Israel is the *kibbutz*. The members of a kibbutz live together and share everything. Instead of being paid wages, they receive their food, clothes and shelter (a flat or a bed-sitting-room).

One of the first and finest houses to be built in a kibbutz is the children's house. Here the children live while their fathers and mothers are working. The children spend the evenings and holidays with their parents, but they go back to the children's house to sleep.

In hot countries, fig trees produce two or three crops each year. The figs sent to Britain are dried in the sun or in ovens

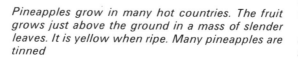

Pineapples grow in many hot countries. The fruit grows just above the ground in a mass of slender leaves. It is yellow when ripe. Many pineapples are tinned

In the orange grove of a kibbutz in northern Israel. The crates and the rails they run on were made in the kibbutz's factory

Some of these settlements have as many as 2000 people living in them. The members have regular meetings to decide all the affairs of the kibbutz, and they elect managers to be in charge of each department—the farm, the factory, the dining hall, the children's house and so on.

The kibbutzim are basically settlements for farming and defending the land, but they all have libraries, club rooms and so on, and many have museums, and factories. In the factories they make anything from simple agricultural equipment to toys and luxury furniture.

The factories

Some of the money needed by the state of Israel is given by Jews in other countries; but the government wants to produce more goods for export, in order to bring more money into the country. The people are

encouraged to set up factories, particularly in remote parts of the country, where use can be made of local raw materials. Potash, salt and other minerals found in the Dead Sea are used in the chemical factories. Minerals such as copper, several types of china clay, and sand for glass-making are found in the Negev, and so Israel has metal, pottery and glass industries. Cotton is made into clothes and locally grown fruit is tinned. The polishing of diamonds is also an important industry. Many of the factories are small, but they are very efficient. Larger factories, built recently, include a steel mill at Acre, and several plants at which cars and trucks are assembled (including a Leyland factory).

Most of the goods imported and exported are shipped through the Mediterranean ports of Ashdod and Haifa, while the Red Sea port of Eilat provides Israel with a route for her trade with the Far East and Africa.

These modern flats in Beersheba have balconies where the residents can enjoy the fresh air while being shaded from the sun

Israel today

Many of the towns are a mixture of old and new. The old parts have narrow lanes and dark, stone houses. The new parts of the towns have been laid out with trees, squares and wide roads. Twenty years ago Beersheba was just a ramshackle market town for nomadic tribes of Arabs. Today it is a thriving modern city, with busy shopping centres, hotels, factories and cinemas.

Beersheba is the southern terminus of the railways. The major towns are linked by railway and more lines are being built. There are many fine asphalted roads. One of the best new roads is from Beersheba southwards to Eilat. In 1948 Israel had 4 ships and 2 civilian planes. Now she has more than 70 ships (most of them cargo ships, used particularly for carrying citrus fruits). "El Al" airways have regular flights to America, Europe and South Africa.

Israel has made great progress since 1948. Tremendous effort by her people, and a great deal of money from other countries, have enabled her to grow most of the food needed by her ever-increasing population. Tourists, too, spend money in Israel. The Israelis have shown how a hot, dry country in the Middle East can have a standard of living similar to that of Western Europe.

The old city of Jerusalem

294

10 Water in the Hot, Dry Lands

Men cannot live, and plants cannot grow, unless they have water. In hot, dry lands, where there is very little rain, the supply of water is a great problem. In lands where the rain falls only at one time of the year, the rain-water must be stored until it is needed. In lands where hardly any rain falls, water must be brought from areas where there is plenty of water.

Irrigation in hot, dry lands

This man is opening a channel so that water can flow to his cotton plants. Without this watering, or *irrigation*, the plants would die.

Opening a channel so that water can flow from an irrigation ditch on to the cotton plants

The photograph was taken in the Sudan, a hot country where there is very little rain, and the water was brought in channels from the River Nile. Irrigation is needed in dry countries such as the Sudan and Egypt, and also in such countries as Southern Spain and Greece where rain falls only during a few months each year, and not when the crops need it.

Old and New

Some of the earliest civilisations in the world grew up around the River Nile in Africa, the River Ganges in India, and the Yangtze-Kiang river in China. No civilisation would have been possible without the water from these rivers to irrigate the land. The rivers rise in mountainous districts, where the rainfall is heavy, and then flow through dry lands to the sea.

water-wheel   sakya   shaduf   treadmill   Archimedes' screw

Some of the old methods of raising water from a river into an irrigation ditch

Egypt: An aerial view of the River Nile. Notice the cultivated land on either side of the river, with desert beyond

The old ways of raising water are still used by a few farmers. But bigger, national irrigation schemes use huge dams, barrages (large weirs), and electric pumps, which can store and move larger quantities of water.

Most modern schemes start with the damming of a river. This makes an artificial reservoir above the dam. Water from this reservoir can be let out through a series of channels on to the land below, as it is required.

Whenever a dam is built, it serves a double purpose, for it is also used to create hydro-electric power.

The River Nile

Less than one inch of rain falls each year in Egypt, and the land is a barren desert, except where it is irrigated from the River Nile. The Nile brings down a great deal of water in the autumn. At one time it used to flood over its banks and so irrigate the land. Very much less water comes down in the spring and summer. The farmers, however, like a supply of water all the year round, so that they can grow two or three crops a year. In summer, particularly, they need water for the cotton, which is their most valuable crop.

By building a number of dams and barrages, the flow of the Nile has been controlled. The greatest dam is at Aswan, over 400 miles (650 km) south of Cairo. This huge dam is $1\frac{1}{2}$ miles ($2\frac{1}{2}$ km) long. Since it was first built it has twice been made higher, so that it holds the water back for 200 miles up river.

Below the dam there are barrages which check the flow of the water and divert it into irrigation channels.

Egypt: date palms growing on land irrigated by water from the Nile

Irrigating citrus groves with water from a dam on the Murrumbidgee River

Part of the main reservoir of the Snowy Mountains scheme

The Murray River basin, Australia

Much of the huge country of Australia is desert. In south-east Australia the rivers Murray, Murrumbidgee and their tributaries flow across dry, but fertile inland plains. Water from the rivers has been used to make the region one of Australia's most important food-producing areas.

Dams have been built in the Snowy Mountains to make six large reservoirs. The largest of these is the Hume Reservoir. Downstream, barrages have been built across the rivers to direct the water into irrigation canals. Sheep, and beef and dairy cattle, are now kept in these parts, and fodder for the animals is grown there. Farther downstream, grapes and oranges are grown.

The Snowy River, which also rises in the Snowy Mountains, runs southwards through land which has sufficient rain, so its water is not needed for irrigation. Work is almost completed on a large scheme to divert the river westwards. As the water falls from the high catchment areas it is used to make hydro-electric power. It then flows into lands where the water is needed for irrigation. Tunnels to carry the water have been dug through the mountains.

West Pakistan

The River Indus and its tributaries all rise in the Himalayas. The land between the tributaries is called the Punjab (or land of five rivers). In spring, when the snow in the mountains melts, the rivers often flood. But in winter there is little water in the rivers as they flow through the desert of Sind to the sea.

297

The Sukkur barrage, on the River Indus

Dams and barrages have been built on some of the rivers, so water can be stored and used for irrigation. The Sukkur barrage across the River Indus is nearly a mile wide. It has 66 gates and 7 canals lead from it; water from these canals irrigates 6 million acres of land. Another barrage, the Kotri barrage, has been built near Hyderabad to irrigate land which otherwise would be desert. Two crops a year, usually rice and wheat, are now grown on this land.

Other ways of getting water for irrigation

In many parts of the world there are good supplies of water underground. In the plain of the River Ganges, holes are drilled to reach this water, and a perforated steel tube is put into each hole. The water which seeps into the tube is drawn from the well by an electrically driven pump.

In parts of Australia, underground water is obtained from great depths. Water trapped between layers of solid rock gradually seeps downwards. If a well is drilled at a place where the well-head is below part of the underground water, then water will flow from the well without needing pumping. This is called an *artesian* well.

In Russia, China, India, West Africa, and wherever progress is being made in the world towards a higher standard of life, a good water supply is one of the first needs. Without it, no real progress is possible.

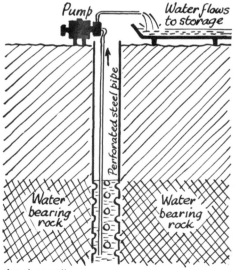

A tube well

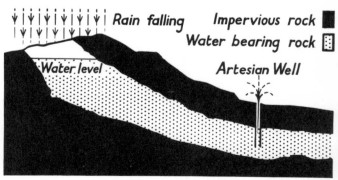

An artesian well

298

Let's remember
The Hot Lands

Many things help to decide the climate of a country. Here are some of them:

1 *The distance from the equator*. Equatorial lands are hot, polar lands are cold.

2 *The height of the land*. The higher the land, the cooler the climate.

3 *Distance from the sea*. Lands near the sea have a less extreme climate than lands in the middle of continents.

If we think about these things, we find that there are several kinds of hot lands with different types of scenery and climate.

1 The equatorial forests

At the equator, where the sun is almost directly overhead all the year round, there is little difference in temperature between one month and another, or between night and day. The temperature is rarely more than 90° F (32° C), but there are thunderstorms and heavy rain almost every afternoon.

In the dense tropical jungle which grows in the lands around the equator, it is damp and

Equatorial forest: Brazil

oppressive. So much rain falls that the ground is often swampy and there are many wide rivers. Tropical forests cover the basin of the River Amazon in South America, of the River Congo in West Africa, and the lands of Malaysia and Indonesia. Few people live in these forests.

2 The tropical grasslands

A little to the north and south of the equator it is very hot and rain falls only in summer. There are wide open spaces covered with coarse grass and a few scattered trees which can withstand the drought of the winter months. Where the tropical grassland, or *savannah*, has been cultivated, crops of coffee (Brazil), cotton (the Sudan), sugar cane or millet are grown. But the main use of the savannah is for rearing cattle.

The savannahs are found north, south and east of the Congo basin in Africa, in the Guianas (where they are called *llanos*), in Central Brazil (called *campos*), and in Northern Australia.

299

Tropical grassland: Botswana. The grass is poor, but cattle can live on it if they move around

A hot desert: the Sahara, Africa

Tropical monsoon land: Ceylon. Wash-day in a mountain stream

3 The tropical monsoon lands

The countries of India, East Pakistan, Burma, Thailand and Indonesia have three distinct seasons each year: the cold season, the hot season, and the hot, wet season.

Where the rainfall is very heavy tropical jungle grows. Many of the trees are deciduous hardwoods such as teak. Where the land has been cleared, rice is the main crop. Other crops are tea (particularly in Assam), jute (the Ganges delta), rubber (Malaysia), cotton and sugar cane.

4 The hot deserts

To the north and south of the savannah lands are deserts which have less than 9 inches of rain a year, and frequently no rain at all. A fierce sun shines from a cloudless sky and very little can grow. In summer it is very hot during the day (110° F or 43° C), and very cold at night. The winters are cooler. When rain falls, many small plants grow quickly and make their seeds, but for most of the year only a few thorny bushes and cacti can survive the drought.

Some of the inhabitants of the desert are nomadic herdsmen; others are oasis-dwellers who grow date palms and keep cattle, sheep and goats. Oil has been found in the deserts of Arabia and in the Sahara Desert, gold in Australia and nitrates in Northern Chile.

The largest area of desert stretches from the Sahara in North Africa across Arabia to Northern India. Other deserts are the Kalahari desert in South Africa, the desert of Central Australia, and the deserts of Northern Chile and of Mexico.

300

An aerial view of Cape Town
with Table Mountain in
the background

THE TEMPERATE LANDS
11 The Republic of South Africa

The land

South Africa is a large country, five times as big as Britain.

Although it is the same distance from the equator as Egypt, it is not nearly so hot, for most of the country is a high plateau, about 3000 ft (900 m) above sea level. Around the coast is a strip of low land.

The story of South Africa

Until the seventeenth century the only people in South Africa were the original natives, the Bushmen and the Hottentots, and the Bantu Africans who had come from farther north. Then the Dutch set up a post where ships trading with India and the Far East could pick up water, vegetables and meat. In the years following, many people from Europe, particularly from Britain and Holland, went to South Africa to settle.

Gradually the settlers moved east and north, until they met the Bantu people, and fighting sometimes broke out for the owner-ship of land. The settlers, who had firearms, conquered the Bantu. Then the settlers quarrelled amongst themselves, and in the Boer War the British fought the Dutch-speaking "Boers".

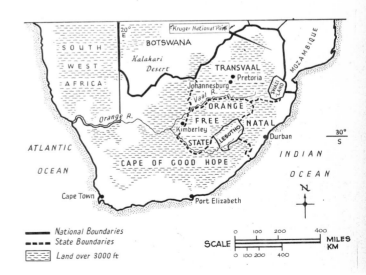

National Boundaries
State Boundaries
Land over 3000 ft

A Bantu miner

A white South African

A "Coloured" businessman

An Indian student

The people

Today, nearly nineteen million people live in the country:

> 12½ million are Bantu.
> Over 3½ million are the descendants of European settlers.
> Over 1½ million are "Coloured" (partly negro, partly European and partly descended from slaves who were brought from the east).
> About ½ million are Asians (some of them are immigrants from India; others are descendants of labourers who came to Natal to work in the sugar-cane fields).

The government of South Africa is in the hands of the "whites". They believe that each race in the country should live a separate life, and should not be mixed with the others, as has happened in Brazil and in other South American countries. They have set aside parts of the country as "reserves" for the Bantu people, and they want the Bantu to keep their own culture and way of life, and to develop their own government with its own leaders. This policy is called *apartheid*, which means "separateness".

Diamonds and gold

Much of the wealth of South Africa has come from under the ground, for many kinds of minerals are mined there. When diamonds and gold were first found in South Africa, almost a hundred years ago, men rushed there from all parts of the world. New townships sprang up almost overnight, and all around Johannesburg men were frantically digging for gold.

The diamond is a very hard stone which can be cut and polished so that it flashes with every beam of light which catches it. It has always been the most prized of jewels. Today, because of its hardness, it is also used in industry, for cutting and grinding, and for making the bearings of watches.

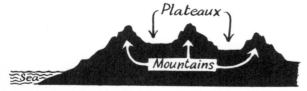
Much of South Africa is a high plateau, or tableland

302

To mine diamonds, the bluish-green rock under the surface of the earth is dug up, crushed and washed so that any diamonds can be picked out. (Many tons of rock are mined to find each diamond.)

In some areas, diamonds lie close to the surface and in river beds. The diamonds are usually exported in their "rough" state to be cut and polished in Amsterdam, Antwerp or Israel.

The diamond mines in South Africa are owned and managed by "whites". They employ skilled and unskilled workmen who are "white", "coloured" and Bantu.

Three-quarters of the world's gold comes from South Africa, most of it from the mines near Johannesburg. Gold comes from deep down in the ground, and huge dumps of bright yellow soil are to be seen all round Johannesburg.

To get an ounce of pure gold, three tons of ore have to be mined, crushed, and treated.

Johannesburg

Uranium and other minerals

Uranium has recently become very important, for it is radio-active and is used in the making of atomic power. It is extracted from the rock which is dug up in gold mining. Uranium is now South Africa's second most important mineral export. Iron ore, manganese, copper and asbestos are also mined in South Africa.

Maize is a grain crop grown in many hot countries. It grows 6–10 ft (2–3 m) tall, with several "cobs" on each plant. In South Africa it is called "mealies" and is very popular with the Bantu people

Farming

South Africa is a land of sunshine, and for months on end there are clear-blue skies and a warm sun.

Fruit grows well, particularly in the south-west of the country, around Cape Town. The summers are hot and dry and the winters are warm and rainy. This sort of climate is often called "Mediterranean". It is the same climate as that of Italy, Greece and California in the U.S.A., which are also famous for their fruit.

South Africa has its summer when the northern hemisphere is having winter, so that its fruit reaches Britain just when it is most needed.

A vineyard in Cape Province

Sugar cane, pineapples, oranges and grape-fruit are grown around Durban and other east coast districts. Around Cape Town, apples, plums, apricots and grapes are grown. South Africa grows a great deal of maize.

Fresh fruit is exported in refrigerated ships. It is chilled but not frozen. Some fruit is tinned or made into jam. Next time you are in a shop, or in a self-service store, look at the labels on the tins of fruit, and see how many come from South Africa. The country is also famous for its wines.

Millions of sheep are kept, mainly in Cape Province and in the Orange Free State. Nearly all the wool is exported.

Animals

Africa is a land of many animals, and at one time it was famous for big-game hunting. Lions, elephants, hippopotamuses and all kinds of animals and birds were hunted, until there was a danger that some of them might become extinct. To safeguard the animals, and to allow people to see them in their natural surroundings, nature reserves were set up. In these areas the hunting of animals is forbidden.

The largest reserve in South Africa is the Kruger National Park, which is larger than the whole of Wales. As you drive through the park, you may see a herd of elephants ambling across the road, a giraffe nibbling the leaves of roadside trees, or a lioness lying asleep on the track, with her cubs playing around her.

Lion

Wildebeest

Cheetah

Elephant

Rhinoceros

Hyena

Baboon

Springbok

Zebra

An old Japanese archway

Workers in a Japanese rice field

12 Japan

The country of Japan is made up of a long string of islands which stretch along the eastern shore of the continent of Asia. There are four main islands and hundreds of smaller ones. The chief cities are on Honshu, which is about the size of Great Britain.

Three-quarters of the country is mountainous, and nearly everyone in the islands lives on the narrow plains which border the sea. Two great cities on these plains are Tokyo, the capital, with its port of Yokohama, and the great industrial centre of Osaka.

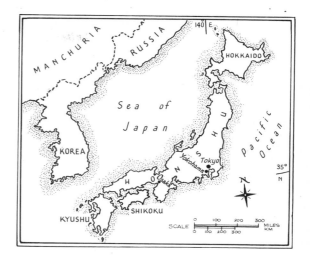

The climate

The islands of Japan stretch so far from north to south that the climate is very varied. Hokkaido, in the north, is a cool, temperate land, while Kyushu, in the south, is much warmer.

Although Hokkaido is nearer to the equator than Britain, it has a colder winter. This is because biting, icy winds sweep across the sea from the north of Russia, bringing snow for four or five months of the year.

The people

Over 100 million people live in Japan, a very large number for such a small country. How do so many people make a living on so little land? The answer is that they work hard, and live simply. Their food is plain, and their houses, by Western standards, are simple. A hundred years ago Japan was a backward country, having no contact with the outside world. Today she is one of the busiest of nations, with factories making cameras, motor scooters, toys and many other things which are sold to other countries.

During the day most Japanese wear European clothes. But on special occasions they wear the traditional dress, including the loose robe called the *kimono*.

The kimono is worn only on special occasions

305

To start a cultured pearl, each oyster has a tiny bead placed inside it

The oysters are put in cages which are tied underneath rafts in the sea

After three years about half the oysters will have made pearls. These are sorted and graded

Fishing

Japan has little good farm land, and to feed her large population cheaply, she depends a great deal on fish. In the villages near the sea, the women work in the small rice patches, while the men earn their living by fishing. Some of the boats work around the long, irregular coast, where the fish are plentiful, but there are also many trawlers which make longer trips. There are large ships which have built-in canning factories from which the fish are sent directly to various countries.

The Japanese live mainly on rice and fish. Here are some of the many kinds of fish which Japan sells to other countries

Frozen foods: tuna, swordfish, rainbow trout, etc.

Tinned foods: crab meat, tuna, sardines, salmon, oysters, etc.

Pearl "farming"

Many people round the coast earn a living by trading in pearls. Pearls are beautiful "stones" which are sometimes found inside shellfish called oysters.

As real pearls are difficult to find, the Japanese grow "cultured" pearls in oysters which they breed in the shallow water close to the shore. This is called pearl farming.

The Textile Industry

For many years Japan has been famous for her silk and cotton cloth. Raw cotton is imported from the U.S.A. and is taken to factories where modern machines, operated by skilled technicians, spin it into yarn. Some of the yarn is exported and the rest is woven into cloth. Spinning and weaving are carried out in the same factory.

Silk cocoons raised by Japanese farmers are sold to factories nearby, where they are reeled by machines. Most of the silk thread is exported but some is woven into fabric.

Nowadays man-made fibres are far more important than cotton and silk and Japan makes large quantities of nylon and other synthetic materials.

Farming

Farm land is scarce in Japan, for much of the land is mountainous and less than one-sixth of it can be farmed. Much of this land is poor, but it is well farmed and good crops are grown. Mr Sato has a farm on the crowded plains of Honshu. Here the farms are very small; each has only about three acres of land. (An acre is about the size of a football pitch.)

Most farms are made up of several scattered plots, instead of having all the fields in one place. Some of these plots are on the

Picking tea from bushes grown on the terraced fields of a hillside. Rice is growing on the plains below

flat plain lands, while others, only a few feet wide, have been made on the slopes of the mountains. They look like giant steps.

The fields are so small and the country so mountainous that the farmers are unable to use tractors or similar big machinery. Most farmers use powered cultivators or

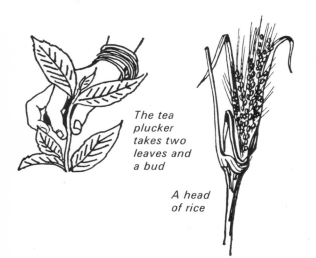

The tea plucker takes two leaves and a bud

A head of rice

hand tractors. Some also keep a few cows, but the main dairy farming area is Hokkaido. The land on the other islands is needed to grow food, and cannot be spared for cattle. Grazing land in Japan is so scarce that butter often costs 50p a pound.

Mr Sato plants two crops a year. Rice is grown in the summer, and after the autumn harvest the land is ploughed again. In winter and early spring the same fields will grow wheat and barley. Because land is so scarce, every inch of the farm is cultivated, and even the banks dividing the fields are planted with tea bushes.

Silkworms feed on the leaves of the mulberry tree, but the Sato family keep fewer silkworms than they did in the past because nowadays many artificial fibres are used instead of pure silk.

Japan

A farmhouse in Honshu

The Satos' farmhouse is made of wood with a thatched roof of bark. The outside walls were made by plastering a special clay on to a framework of thin wood. This was then painted. Sliding screens, called shoji and fusuma, are used for windows and dividing walls. Shoji are very fine screens which shut out the glare of the sunlight but let in a soft light. Fusuma are covered with thick paper and shut out all light.

Inside, the house is simply furnished. The floor is covered with tatami mats made of rice straw. They are two inches thick and very comfortable to sit on. There are no beds, and at night thick, padded quilts are laid on the rush mats which cover the floors.

A Japanese girl writing a letter. She uses a brush and paint to "draw" the words. Here are two Japanese words:

Rice   お米   お茶   Tea

Japan's imports and exports

To feed all her people, Japan has to buy from other countries foodstuffs such as sugar, wheat and soya beans. In order to pay for this food she must manufacture goods and export them.

The country has few raw materials. She has to import petroleum, bauxite for making aluminium, and iron ore and coking coal for her iron and steel industry. She also buys wool from Australia and cotton from the U.S.A.

Japan is the world's greatest shipbuilding nation

Sewing Machines   Bicycles   China   Cameras   Toys   Tinned Mandarin Oranges

Some of the many things exported by Japan

Japan has some coal and copper, pine trees for the match industry, and clay for making china and earthenware. Hydro-electricity from the mountains provides power for many of the factories.

Metal and Machinery

Japan exports millions of tons of pig-iron and crude steel to countries all over the world. Raw materials are imported; iron ore from Malaya and India, coking coal from Australia and scrap iron from the U.S.A. Some of the metal is used to make machinery, cars and ships.

For many years Japan has been the world's biggest shipbuilding nation. Most of the ships are sold to other countries.

Japanese motor-cycles, such as the *Honda*, have been famous all over the world for many years, and some countries also buy buses and lorries from Japan. Now cars are being made.

The Japanese are good at making very small electrical things and their T.V. sets, transistor radios and refrigerators are cheaper than any others in the world.

Railways

The Japanese have built hundreds of fine new locomotives, passenger coaches and goods trucks and their railways are fast and efficient. Many countries now buy rolling-stock from Japan.

Earthquakes

An earthquake is a shaking of the ground caused by violent movements far beneath the earth. Japan has more earthquakes than any other country. Almost every day, throughout the islands, there is a small earthquake somewhere. During these earth tremors, the houses shake, the earth trembles for perhaps half a minute, and often crockery is broken.

In 1923 the worst earthquake in Japan's history destroyed most of Tokyo and Yokohama, and both cities had to be rebuilt.

When these cars are completed, many of them will be exported to other countries

13 A Train Journey Across Europe

from London to Venice

From London to Paris

Mr. Taylor is a "buyer" for a London Store. He is travelling by train from London to Venice, a journey which will take him through France, Switzerland and Italy.

He decides to cross to France from Newhaven to Dieppe. Here is the timetable of the journey as far as Berne. How many miles is it to Paris, and how long will it take him to get there? (Remember that continental timetables use a "24-hour" clock.)

Mr. Taylor's route from London to Venice

The port of Dieppe. The large boat is a cross-Channel steamer

Time	LONDON—PARIS	Miles
8.50	*dep.* London (Victoria)	0
10.05 *arr.* 10.45 *dep.* } Newhaven Harbour		56
15.0 *arr.* 15.30 *dep.* } Dieppe Maritime		130
16.30	*arr.* Rouen	169
18.08	*arr.* Paris	235
	PARIS—BERNE	
8.10	*dep.* Paris*	0
13.0	*arr.* Frasne	274
13.26	*arr.* Pontarlier†	284
14.44	*arr.* Neuchâtel	317
15.46	*arr.* Berne	387

* Restaurant Car
† Frontier station for customs, passport examination and exchange of money

Mr. Taylor crosses the Channel on a British Railways steamer. The first part of his journey across France is through Normandy. The orchards of cider apples, and the market gardens and dairy farms of this fertile province, provide much food for Paris.

After an hour in the train, Mr. Taylor sees the slate roofs of Rouen, an important industrial town on the banks of the River Seine. Rouen is also a busy seaport, since the Seine has been dredged to enable cargo boats to reach the city.

Rouen is a busy port even though it is 50 miles (80 km) from the sea

The train has a comfortable dining car

In this part of France there are many canals, linking the Seine and its tributaries with many of the other rivers of France, including the Loire, the Meuse and the Rhône. Barges on the canals and rivers of France carry more freight than the railways. Some of the waterways link the north of France with Holland and Belgium.

Punctually at 6.8 p.m. the train arrives in Paris, where Mr. Taylor breaks his journey and spends the night. As he crosses the city by taxi on the way to his hotel, he see brilliantly lit shop windows and busy cafés.

From Paris to Berne

Next morning Mr. Taylor boards the train for Berne. It is a diesel train with a very good restaurant car. Mr. Taylor has breakfast as the train approaches the farming and vine growing provinces of Champagne and Burgundy, where two of the most famous French wines are made.

311

A train journey across Europe

Soon the train climbs up the western slopes of the wooded Jura mountains, on the borders of France and Switzerland. The thick forests provide wood for sawmills and furniture factories, and many of the towns and villages have yards beside the railway, where lengths of timber are stacked in piles ready for transport to other parts of France. The hard wood of the box tree, which grows in the mountains, is made into wooden toys, many of them carved by hand.

After the train has climbed above the forests, the country opens out into pasture land. Horses, sheep and goats are reared on the grassy uplands of the Juras. In winter, the weather is cold and damp and there are heavy falls of snow. In the worst weather, the animals are kept in large barns which have ramps leading up to them.

At the frontier between France and Switzerland, customs officials come along the train, as it moves along, to examine the passengers' luggage and to stamp their passports.

Avalanche shelters are built over the line to protect it from falling snow and rock

Then the train continues its long descent down the eastern side of the Jura mountains from the Swiss frontier. Here there are many rivers flowing towards the lakes of Geneva and Neuchâtel. The railway goes through tunnels and cuttings in the rock, and as the train enters each tunnel the engine driver blows a horn.

The people of the Swiss valleys of the Jura mountains are famous for the manufacture of things which need skilled and accurate workmanship: watches, clocks, cameras and microscopes.

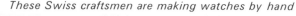

These Swiss craftsmen are making watches by hand

This Swiss watch shows the date, and the time in two different zones of the world

From Berne to Milan

Beyond Berne the train goes through some of the finest mountain scenery in Switzerland. It runs alongside the Lake of Thun, and then begins its climb up the Kander valley to the winter sports town of Kandersteg. This part is called the Bernese Oberland.

Many Jersey cows graze in the rich meadows, but above the meadows, where the mountains are higher and steeper and the soil is poor, there is often nothing but rock and pine trees. Some of the little towns have factories where the pine wood is made into matches.

In the Kander valley, as in hundreds of other mountain valleys all over Switzerland, many people work at dairy farms and factories, making cheese, condensed milk and butter.

Twice a day, in the early morning and evening, the farmers who live on the mountain slopes send their milk to the cheese factory in the valley. Sometimes children

Near the winter sports centre of Kandersteg

bring the churns down the mountain tracks in little two-wheeled carts pulled by huge dogs.

Storing the cheese
The cheeses are stored in cellars for 6–10 weeks to mature. They are washed and turned twice a week—a mammoth task as each cheese weighs about 175 lb (80 kg)

Weighing the milk at the cheese factory

313

A train journey across Europe

In the early summer, the cows are driven to the upland pastures where they remain until the autumn. The lower meadows are then sown with grass and clover, which is later cut and stored ready for winter fodder when the ground is covered with snow.

During the summer the cattle live on the high mountain pastures

On the upland pastures there are little huts built of stone, with roofs of wooden shingles. Some of the people live in these huts during the summer months. They milk the cows and make cheese.

The train goes along the mountain slopes to the valley of the River Rhône

In the past, when the cows were on the upland slopes, the villages in the valleys were often short of milk. Today, in some parts of Switzerland, a new idea has been introduced from Austria. Polythene pipelines are buried, about two feet under the soil, so that milk can be sent down the mountain to a collecting "station" in the valley.

The train crosses the Bernese Alps by the Lotschberg Tunnel, and then winds down the slopes of the mountain to the valley of the Rhône, the largest valley in Switzerland. This valley, which is protected by the Bernese Alps on one side, and the Pennine Alps on the other, is an important area for growing fruit and tomatoes. Not an inch of its fertile soil is wasted.

In the Rhône valley, as in many of the valleys of Switzerland, there are huge hydro-electric stations. Pipelines down the mountain sides carry the water from the mountain lakes to the generating stations in the valleys. The lower slopes of the mountains, and the spaces between the fruit trees, are planted with grape vines. Some varieties of grapes are for eating; others are made into wine.

Before the train reaches the frontier town of Brigue, a money changer comes along the train, and Mr. Taylor changes his French and Swiss *francs* into Italian *lire*.

At Brigue, the train enters the Simplon Tunnel. This tunnel, $12\frac{1}{4}$ miles (20 km) long, took seven years to build. The frontier between Switzerland and Italy is in the tunnel.

At the town of Domodossola over the frontier the Swiss engine is changed

A money changer comes along the train

for an Italian engine. Two hours later the train reaches Milan, the most important industrial city of north Italy.

As Mr. Taylor comes out of the vast station he sees many new shops and office blocks. Italy is famous for good design, in buildings, clothes, furniture, cars and machinery.

Milan

Mr. Taylor spends several days in Milan. He visits the Pirelli factory to place orders for sports-shoes and foam-rubber mattresses. Tyres and many other kinds of rubber and plastic articles are also made there. Mr. Taylor later visits a silk factory and orders silk scarves for his store. Italian silk fabrics have been famous for centuries, although today, now that so many "man-made" fabrics such as nylon are being worn, only about one mill in ten produces pure silk.

A new office block in Milan

315

In Northern Italy there are many factories making cars, cycles and motor-scooters. On a track near Milan, racing cars from many European countries compete against Italy's Ferrari and Maserati cars. Alfa Romeo cars are made in Milan; Ferrari and Maserati cars are made in Modena.

After leaving Milan, Mr. Taylor travels on by train eastwards across the broad Plain of Lombardy. In many parts of Italy, farming is difficult because of the shortage of water. The Plain of Lombardy, however, has plenty of water. The River Po crosses the plain from west to east; tributaries from the Alps in the north, and from the Apennine mountains in the south, flow across the plain to join the main river.

Two of the most important crops grown on the Lombardy Plain are rice and maize. Instead of hedges, rows of mulberry bushes, or grape vines trained over low wooden frameworks, divide the fields. In autumn,

A farmhouse on the Lombardy Plain. The farmer is hanging up maize cobs to dry

the farmers and their families harvest the tall maize plants, storing the stalks in barns for use as bedding for the cattle. The bright orange maize cobs are dried on the farmyard floor, or under the eaves and balconies. Later they are ground into flour.

Venice

At last the train reaches Venice, one of the most unusual cities in the world, for it is built on over a hundred islands in the middle of a lagoon. The houses are built on wooden piles driven into the mud, and the islands are linked by bridges across the canals.

Venice is built on over a hundred islands. The railway line and road are carried over the lagoon by a bridge $2\frac{1}{4}$ miles long

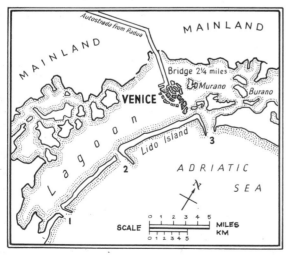

A Venetian gondola

The Grand Canal is the "High Street" of Venice

Mr. Taylor walks down the steps from the railway station to a quay on the Grand Canal, the main waterway of Venice. He takes a water-bus to his hotel, for no cars can use the narrow pathways of Venice. Everything in Venice goes by boat: firewood, bags of cement—even the household rubbish.

Before Mr. Taylor returns to London, he visits one of the glass factories on the island of Murano. Here, for over 600 years, the beautiful Venetian glass has been made.

In summer, the workers start at dawn because of the intense heat of the furnaces. Inside these furnaces there are fireproof tanks filled with molten glass of different colours. The men dip long iron bars into the "treacly" mixture, and shape the hot glass with pincers, scissors and other simple tools.

When each article is finished, it is left to cool for 48 hours, and then carefully packed ready for export.

At the end of his tour Mr. Taylor flies back to London. On the way, he reckons up the value of his trip. It has cost his firm a good deal of money, but it has been well spent, for he has ordered many exciting things which should sell well in his store at home, and he has gained many new ideas from seeing the stores of Paris, Milan and Venice.

Making a glass bird on the island of Murano

The "Stars and Stripes", the flag of the U.S.A. The stripes represent the 13 original states of the Union. The stars represent the present 50 states

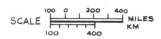

SCALE

The 48 mainland states of the United States of America

14 The North-east of the U.S.A.

In Chapter 6 you read about the Southern States of the United States of America. In this chapter you will read about the north-east part of this enormous country. But first let us take a quick look at the 48 states which make up "continental" U.S.A. (Alaska and Hawaii are the other two states of the Union.)

The 48 States of the Union

The main body of the U.S.A. is a wide belt of land which stretches from the Atlantic Ocean in the east to the Pacific Ocean in the west, and from Canada in the north to Mexico and the Gulf of Mexico in the south. The country is so large that an express train, travelling at a mile a minute (95 km an hour), takes 48 hours to cross it.

In the north-east, the states of New York and Pennsylvania have busy industrial areas and many large cities. A

range of mountains, the Appalachians, separates the Atlantic coastal states from the vast plains of Illinois, Iowa, Nebraska and Arkansas. Much wheat and corn are grown on these plains, and the great River Mississippi flows through them southwards to the sea.

In the west are the Rockies, the great range of mountains running from Canada in the north, to the state of New Mexico in the south. In Arizona, California and in Utah, there are vast deserts. The climate of the west coast ranges from the damp, cool north (rather like Britain), where there are large forests, to the dry, hot climate of the south. Fruit is grown all along the west coast, but particularly in California. Films are also made there, for the air is so clear that filming out-of-doors is possible on nearly every day of the year.

As well as having very varied scenery and climates, the states vary in many other ways. For example, they have separate governments, laws and school systems. Many have people of different races and nationalities living in them. In some of the Southern States, for example, there are as many negroes as white people. In Cleveland, Ohio, there are more Hungarians than in any other city in the world except Budapest.

This sports stadium in Wisconsin seats over 32 000 people, and has parking space for 7000 cars

Yet despite all these important differences, the states of America, and the people, have been welded together to make a united country, with a people whose first loyalty is to their country. The U.S.A. is today one of the wealthiest countries in the world. It produces more of the following than any other country:

> *minerals:* petroleum, copper, lead, zinc
> *crops:* cotton, maize   *metals:* steel
> *"man-made" goods:* rayon, synthetic rubber

As a result, many of the people have cars, telephones, refrigerators and similar luxuries.

The north-east of the U.S.A.
Plentiful coal and iron have made the north-east the most important industrial area of the country, and today a network of up-to-date canals, roads, railways, and air services links the many factories, towns and ports.

The factories make every kind of manufactured goods, both for use in the U.S.A. and for export, particularly across the North Atlantic Ocean to Europe. Much of the trade is through the ports of Boston, New York, Philadelphia and Baltimore. (See the map on page 322.)

The "cloverleaf" enables cars to enter, leave or cross a main road without stopping

New York

New York is a city of eight million people. It is not the capital of the U.S.A., but it is the biggest and busiest of her cities.

The heart of the city is on the island of Manhattan, between the mouth of the Hudson River and the East River. These rivers are deep, and large liners can sail up them to the heart of the city. Because New York is linked by the Erie Canal to the Great Lakes, it has become the U.S.A.'s busiest port. It is also a very busy airline centre.

The centre of New York is on an island, so there is little space for growth. Fortunately Manhattan is an island of solid rock, which provides a firm foundation for tall buildings. As a result, the city has grown upwards as well as outwards, and there are many skyscrapers. The tallest, the Empire State Building, has 102 storeys and is 1472 ft (430 m) high.

The plan of New York is very simple: avenues run north and south; streets run east and west. This makes it much easier to trace an address in New York than in London, or most other cities. British visitors to New York usually find the hotels and offices too hot for them, in spite of air-conditioning. Americans like a good central heating system, and they need it in the New York winter, when it is bitterly cold. In summer, temperatures are much higher than in London.

Many of the people who work in the banks, offices and shops of the city travel home each night to the suburbs of New York, which sprawl for miles on the mainland.

New York—the skyline of Manhattan at night, seen under the Brooklyn Bridge

Most of the people who work in Pittsburgh live in the suburbs, on the heights around the city

Pittsburgh

The picture above shows the great steel-making city of Pittsburgh. This huge city is criss-crossed by a maze of railways, roads and canals which carry iron ore, coal and limestone to the steel mills, and finished steel from the mills to other parts of the U.S.A. Pittsburgh is one of the biggest cities in the industrial north-east. It produces one-fifth of America's steel.

The iron ore from which the steel is made is quarried to the south and west of Lake Superior, and loaded into railway wagons which carry it to ports on the Great Lakes. Barges take the ore to Chicago, Detroit and Cleveland.

The coal comes from the Alleghany coal-field, around Pittsburgh. Iron ore can be carried more cheaply than coal, so the steel industry has grown up on the coal-field. Alleghany coal is easy to mine, because the seams are thick and level, and because river valleys have cut through the coal seams. Miners there have only to tunnel into the valley sides to reach the coal, instead of having to dig deep pits, as is usually necessary in Britain. The coal is good coking coal for steel making.

Engineers testing new mining methods

321

The Great Lakes and the St. Lawrence Seaway

There are many important ports on the Great Lakes. From these ports ships carry wheat, iron ore, steel, and many other cargoes, taking them to other ports in Canada and the U.S.A., and to countries overseas.

The locks at Sault Sainte Marie

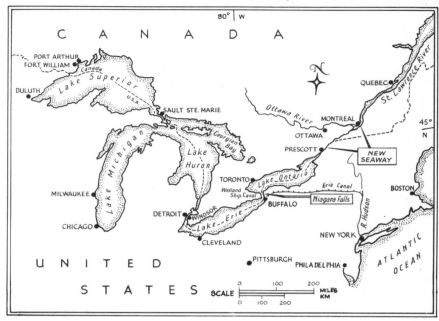

The Great Lakes and the St. Lawrence Seaway

Until recently, large sea-going ships could not sail all the way up the St. Lawrence River to the Great Lakes because of rapids in the river. But in 1959 the St. Lawrence Seaway was opened. This is a series of canals between Montreal and Lake Ontario which by-pass the rapids. To make the seaway, new canals were dug, old canals were deepened and widened, bridges were raised, and locks were made. Large ships can now sail all the way to Chicago and Duluth—but only for eight months of the year. During the four months of winter the lakes are frozen over.

A Great Lakes iron-ore carrier. The ship is divided into several compartments

A small farm in Vermont, New England

New England

The states north of New York are called New England, because many Englishmen were amongst the first settlers there. Vast forests of coniferous trees grow on the mountain sides; in the lowlands, by the coast, farming is difficult because the soil is poor and stony and the winters are very long and cold. The farms produce food for the industrial towns to the south: potatoes, fruit, milk, butter and cheese. Many farmers work only part-time on the land; they also work at lumbering in the forests, and they fish for cod, herring and lobsters in the sea, and for trout and salmon in the rivers.

Timber from the forests is made into furniture and paper at factories in New England. Most other factories there have to import their raw materials: they make cotton, woollen and linen cloth, shoes and metal goods. Ships are built in Boston, the port for New England.

Many Americans visit New England for their holiday. In summer they ramble through the forests and hills; in winter they ski on the mountain slopes.

Cutting down a tree with a power-driven saw

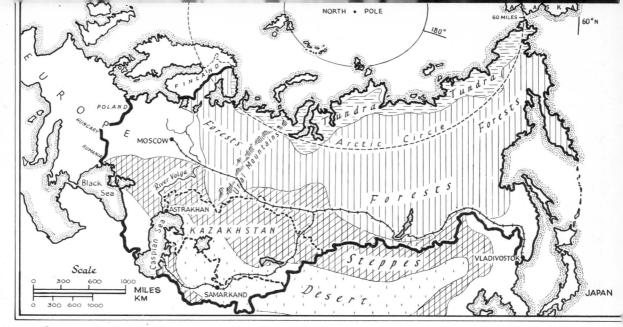

The bold black line shows the border of the Soviet Union

15 Russia

Russia, the largest country in the world, stretches from Eastern Europe right across Asia to the borders of China. The correct name of this vast country is the U.S.S.R., or Union of Soviet Socialist Republics. Russia is really the name of the largest and most important of the fifteen republics, each of which has its own capital, but in this chapter "Russia" means the Soviet Union. Moscow is the capital of the republic of Russia and of the U.S.S.R.

During the last forty years Russia has changed from a backward country with few factories into one of the greatest industrial nations of the world. The country is rich in natural resources: there are plentiful supplies of water power for making hydro-electricity, uranium for making atomic power, oil, coal, iron ore and other minerals. Wool, cotton, grain and timber are produced in great quantity.

Many young Russians have been trained as engineers and scientists, and as a result the Russians are able to build fine tractors, lorries and trains, aeroplanes which can fly non-stop from Moscow to New York, and rockets which can travel in outer space.

324   *Yuri Gagarin, the first Russian astronaut*

The largest country in the world

Russia is so vast, that some of the western cities are as far from those in the east, as Cape Town is from London. In order that people may travel to the other side of their country, a great railway has been built, with many branch lines. The main line of this railway runs from Moscow in the west to the port of Vladivostok in the east, a distance of over 4000 miles (6500 km). The passengers eat and sleep on the train, and the journey takes about a week.

Because of the huge size of Russia, there are great differences in climate between various parts. Much of Russia is far from the sea, and many places have long cold winters. In Moscow, the temperature in January is only 12° F (20° F *below* freezing point——11°C). In the Black Sea area, at Yalta, the January temperature is 38° F, 3° C, and the summers are hot and dry. This part is sometimes called the Russian Riviera.

A family in Moscow

Boris and Olga Beglov, with their children Katya and Sasha, live in a new block of flats in the centre of Moscow. The rooms have double glass windows, one in front of the other, to keep out the cold, and in the winter the flat is centrally heated. If any room becomes overheated, the ventilators in the windows are opened for a few minutes to cool the air.

When the Beglov family go out-of-doors in winter, they wear thick clothing to keep themselves warm. Every morning, gangs of men and women armed with crowbars break the ice from the pavements, and mechanical grabs tip the snow down manholes into the city's underground streams. After a heavy snowfall, special scrapers, each joined to a moving belt, are brought into the streets. The snow is scooped on to these belts, and taken away by fleets of lorries.

Winter in Moscow. The mechanical grab is tipping snow down a manhole into an underground stream

On a winter's day the sky is often bright and cloudless, and the air is clean and dry. The city is lit and heated mainly by electricity or natural gas. (Natural gas is found gushing out of rocks, usually where there are oilfields. Most of Moscow's gas supply is brought over 800 miles (1300 km) by pipelines.) The only smoke in the city is from an occasional power station; there are no coal fires, and steam trains do not come into the city.

Work and play

Like many Russian women, Mrs. Beglov has a full-time job. She works in a motor-car factory where her husband is a fore-man. The factory is owned by the state, as are all the factories, mines, forests, banks and railways in Russia.

The Beglovs travel to work on the underground railway. Moscow's underground stations are almost like palaces. Many of them are built of white marble brought from the Ural mountains.

Each station is different; the walls are decorated with paintings and marble statues, and magnificent chandeliers hang from the roofs.

Saturday is a short working day, and Sunday is the main holiday of the week. Cafés and stores are open, and the ballet, the puppet theatre and the circus are always crowded.

In summer, the temperature of Moscow is higher than that of southern England. People swim in the park lakes, and water buses travel along the River Moskva. The heat in the city is so fierce that the Beglovs, like many other Moscow workers, rent a small log cabin in the pinewood suburbs, outside Moscow, and travel to work by electric train.

Most Russian schoolboys wear a dark blue uniform with a cap

Gorky Street, one of Moscow's main streets

A woman traffic director, with baton and whistle

A school library

School for the younger children finishes in May, and does not start again until the autumn. During the summer, while their parents are working, Katya and Sasha live at a camp for the children of motor-car workers.

Farming in Russia

If you look at the map on page 324, you will see that in the far north of Russia are the frozen lands of the *tundra*, and to the south of the tundra are the great coniferous forests. South of the forest there are grassy plains called *steppes*. Much of this grassland is rich farming country, where grain is grown and big herds of dairy cattle are reared.

All land in Russia belongs to the state, and the people work for the state instead of having farms of their own. Most of the small farms in each district have been joined together to make large farms, known as "collectives". The collective farm is really a village with the old boundary hedges removed and with its own school, library and hospital.

Ploughing, planting and harvesting must be done in the five spring and summer months—from May (when the snow melts), to the end of September. As most of the farms are large, a great deal of machinery is used.

Harvest is the busiest time on the collective farm. Fields of ripening wheat stretch endlessly into the distance, and on the poorer soil rye and barley are cultivated. Everywhere there are patches of sunflowers, whose stalks, with their huge flower heads, grow taller than the harvesters.

The whole village helps during the harvest; even the children and old people. The workers have their midday meal in the communal dining room, or hot dinners are taken to them in the fields by a travelling kitchen.

Harvesting grain on the steppes at Kazakhstan

After the harvest, the government buys the crops, using some of the profits to buy new farm machinery. The rest of the money is shared between the farmers, everyone being paid according to the number of days he has worked and how well he has worked. Each year there are prizes for the best workers: the prize may be a week in Moscow or a free winter holiday by the Black Sea.

Most of the workers have a little land of their own, and they may do what they like with the fruit and vegetables which they grow. Sometimes, on his free day, a farm worker may take his surplus produce on a farm cart to the collective farm market where he sells it at controlled prices.

In the country the workers buy very little food, as nearly all the food they need is produced in their gardens or on the farm. They are given meat, eggs and milk as part of their wages. Their "black bread" is made from rye grown in the fields.

During the long winter months, when the ground is covered over with snow, there are no fresh vegetables. Instead the people eat bottled tomatoes, cabbages which have been shredded and preserved between layers of salt, and short, fat cucumbers called gherkins, pickled in jars of salt water.

The dry steppelands

Farther south, in the Soviet republic of Kazakhstan, the steppes are being newly ploughed for grain-growing in territory called "the virgin lands". The people also keep flocks of sheep. Nowadays most Kazakhs work on state or collective farms producing grain or livestock.

Comfortable houses built for new settlers in Kazakhstan

The cotton harvest

The desert lands

The hot, dry Soviet Republics of Central Asia stretch from the eastern shore of the Caspian Sea to the borders of Persia and Afghanistan. Here, as well as great deserts, there are high mountain ranges, whose snows for centuries have provided the people with water to irrigate the land. Nowadays, more water is needed, and by changing the course of the rivers, Russian engineers are building a network of pipelines and irrigation canals. As soon as water has made the barren soil fertile, new vineyards, apricot orchards and cotton fields are planted.

Many people of these dry lands are Muslims and for centuries the cities which have grown up round their oases have been important trading centres. The most famous of these eastern cities is Samarkand, where Chinese traders once brought caravans of camels and donkeys laden with the spices, pearls and silks of the East.

Nowadays, a railway line joins the cities, and jet planes fly over the old caravan routes.

Right: *A Central Asian family. This part of Russia is famous for its woven carpets*

Below: *Many Russian households have a samovar, which is heated by charcoal or electricity and holds hot water for tea making.*

329

16 Some Countries of
South America

From north to south, the continent of South America is nearly 5000 miles (8000 km) in length. In this great distance there are many different kinds of climate and vegetation, as you can see from the map.

Round the equator, where there is heavy rain on most days of the year, are the hot steaming forests of the Amazon.

In Brazil, Uruguay and Argentina are hundreds of miles of grassy plains. These grasslands grow wheat, and also provide food for large herds of cattle.

The southern part of the continent is much cooler than the north, because it is farther from the equator. Throughout the year, westerly rain-carrying winds blow across the Pacific Ocean, but the *Andes* mountains keep off the rain from southern Argentina. This means that southern Argentina, called Patagonia, has a dry, almost desert climate, in which little grows except shrubs and coarse grasses. It is too dry for cattle, but in some parts sheep are kept.

Parting the coat of a Patagonian sheep to show its thick, rich wool

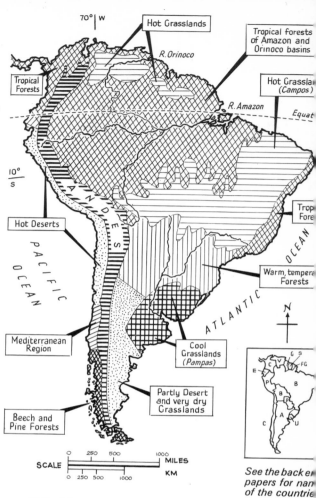

Tropical forests of Amazon and Orinoco basins

Hot Grasslands

R. Orinoco

70° W

Tropical Forests

Hot Grassla (Campos)

R. Amazon

Equat

10° S

ANDES

Hot Deserts

Trop Fore

PACIFIC OCEAN

Warm, tempera Forests

ATLANTIC OCEAN

N

Mediterranean Region

Cool Grasslands (Pampas)

Partly Desert and very dry Grasslands

Beech and Pine Forests

SCALE

0   250   500   1000   MILES

0   250 500   1000   KM

See the back e papers for nam of the countrie

Chile

Southern Chile

On the *western* side of the Andes, stretching from the tropics to Cape Horn, is the long, narrow country of Chile. Southern Chile has a cold damp climate with a heavy rainfall brought by the westerly winds. Here, the coastline is cut into steep-sided fiords covered with forests.

Central Chile

Travelling northwards through Chile, the climate gradually becomes warmer and the forests disappear. Central Chile, which has

330

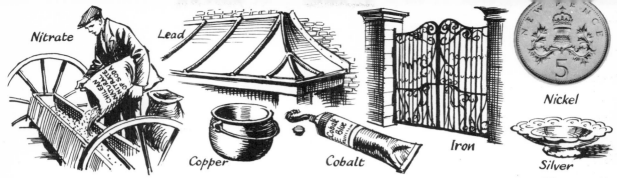

Nitrate   Lead   Iron   Nickel

Copper   Cobalt   Silver

Some of the many uses of the minerals mined in Chile

a climate like that of the Mediterranean lands (warm, wet winters and hot, dry summers), is the most fertile part of the country. The hot summers ripen fruits such as oranges, lemons, apricots and melons. Wine made from Chilean grapes is exported to Europe and the United States.

Northern Chile

To the north of this fertile land is the waterless, sunbaked, Atacama Desert of northern Chile. Nitrate, which is used by farmers as an artificial manure, is obtained from rocks in the desert. (Copper, iron ore and nitrate are Chile's most important exports.)

When the nitrate leaves the factory, as a white powder, it is taken by rail to Antofagasta and other ports on the nitrate coast. Ships sail from these ports to all parts of the world, loaded with nitrate, copper and cargoes of tin from the mines of Bolivia. (Bolivia has no coastline of her own.)

Nothing will grow in the Atacama Desert, so fruit and vegetables for the miners are flown from Central Chile to the ports of the coast, and water is brought by pipelines from the Andes. The water for Antofagasta is piped for nearly 200 miles (300 km).

Over the Andes

The Andes (which, after the Himalayas, is the world's highest mountain range) stretches from north to south through the countries on the western side of South America. Even at the equator the peaks are so high that they are always covered with snow.

Nitrate mining. Gunpowder is poured into drill holes and then exploded to loosen the surface rock of the desert

A herd of llamas in the Andes

331

An Indian house. The woman is spinning llama wool

Indian boats on Lake Titicaca. As the lake is above the height at which trees will grow, the boats are made of rushes

The Andes is not a single range of mountains, but a number of ranges with plateaus (plains) between. There are few roads and railways in the high Andes, and it is difficult and costly to build them.

The railway from the sea coast of Chile to La Paz, the largest city in Bolivia, climbs to a height of 15 000 ft (4600 m) (over three times as high as Ben Nevis, the highest peak in Britain).

The Indians of the high plateau

Much of the plateau country is sandy, stony desert or poor grassland. The Indians breed sheep, and look after flocks of goats, llamas and alpacas. Both the llama and the alpaca are members of the camel family. The llama provides milk and can carry loads of over a hundred pounds. The alpaca is reared for its long, silky wool.

Many of the towns on the plateau have been built where minerals are mined. Railways link the mines with towns on the Pacific coast. Visitors to the plateau find it very difficult to breathe because of the shortage of oxygen in the air, for the higher one goes the "thinner" is the air. But the Indians who work in the mines have larger lungs than Europeans, and as they are used to the "thin" atmosphere, they can work hard without strain.

An Indian house, built of sun-dried mud bricks, often has only one room. As the house is unheated it has solid walls, with only one small window. The floor is made of beaten earth, and the roof of grass or barley straw.

The Indian women spin sheep's wool or the coarser llama wool, and weave it into blankets and clothing. Some of their textiles are very beautiful, and many Indians still wear their traditional dress, even though market stalls in the towns sell ready-made clothing in American styles.

Hereford cattle, originally imported from England, being rounded up on the pampas

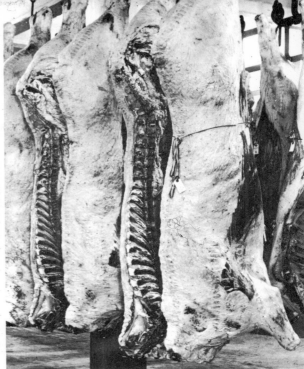

Beef carcases hanging in a refrigeration plant before being shipped to other countries

Argentina

The original inhabitants of South America were the American Indians. Then, in the sixteenth and later centuries, people came from other countries, mostly from Europe, and particularly from Spain, Portugal and Italy. Today in South America, Indians and the descendants of Europeans are mixed throughout the land. But in Argentina there are hardly any Indians, and nearly all the people are descended from Europeans. From the grassy plains called the *pampas*, they have made a wealthy country where they rear cattle and grow wheat.

The pampas was not very useful until five things happened:

1 *Good cattle* were brought from Europe.

2 *People* came from Europe to work the land and look after the cattle.

3 *Wire fences* were put up to prevent the cattle from straying.

4 *Railways* were built to carry the cattle and grain to the ports. (It was easy to build railways on the level land of the pampas.)

5 *Refrigeration* was invented, which meant that meat could be kept fresh on the long sea voyage to other countries.

Buenos Aires is the largest city in Argentina, and the ninth largest city in the world. Argentina imports steel goods, machinery, fuel oil and textiles and pays for them by selling her meat, hides, wool and wheat.

Maori children at school

17 New Zealand

The Maoris

A Dutch navigator called Tasman discovered the islands of New Zealand, naming them New Zealand after the Dutch province of Zeeland. Later, Captain Cook explored the coasts and sailed through the 10-mile wide strait dividing the North and South Islands.

Cook found an intelligent, brown-skinned people called *Maoris* living in New Zealand. About four hundred years earlier these people had travelled southwards from the Pacific Islands in search of a new home. As they were used to the heat of the tropics they settled mainly in the North Island, which is nearer to the equator, and is the warmer of the two islands.

The Maoris lived in fortified villages called *pas*, and gave the settlements names in their own language, like Wanganui (great harbour). In his journal Cook describes their customs, their wonderfully carved houses, and their clothes made from flax.

Settlers from Britain

Apart from small settlements of traders and sealers, it was not until 1840, just over sixty years after Cook's last visit, that people from Britain began to emigrate to New Zealand. They named their settlements after British towns — such as Canterbury, Cambridge, Hastings and Stratford.

Today there are fourteen times more Europeans than Maoris, and most Maoris speak both their own language and English. The Dominion has no "colour bar", and a Maori can marry a European. There are Maori Members of Parliament, lawyers and doctors—in fact, the Maoris have equal rights with the white New Zealanders.

New Zealand needs more British immigrants. The Dominion is nearly as big as the British Isles, and yet in Greater London alone there are four times as many people as there are in the whole of New Zealand.

The kiwi is a bird found only in New Zealand. It has very tiny wings and cannot fly, but it can run very quickly

334

The country which the settlers found was very varied. The South Island has a high ridge of mountains called the Southern Alps running down the west side. The North Island has mountains in the centre. Nearly all the British immigrants took up farming, settling first on the plains near the coast.

North Island

Most of the country's dairy farms are on the coastal plains of the North Island. Here, as rain falls throughout the year, there is always good grassland, and many Jersey cattle are kept. The winter is warm enough for the cattle to be turned out to pasture, and so a New Zealand dairy farm does not need cowsheds. In some places, where the winters are cooler, the cattle wear canvas covers for the winter months.

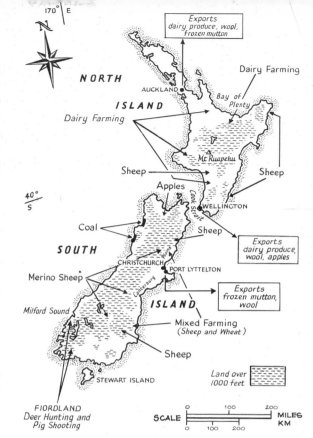

A cow wearing its winter cover

Most farmers use milking machines which are driven by hydro-electric power brought to the farm by overhead wires. The dairy produce of the North Island is exported from the seaports of Auckland, the largest city, and Wellington, the capital.

Mount Egmont, North Island

335

Eglinton Valley, South Island. Notice the "tree line", above which trees cannot grow

Sheep droving in the Tasman Valley, South Island

South Island

The main winds in the South Island are rain-bearing westerly winds, and the west and south-west coasts have a very heavy rainfall. When these winds reach the mountain barrier of the Southern Alps, the water contained in them falls as rain.

On the mountain slopes on the opposite side of the Southern Alps, and on the Canterbury Plains which lie at their feet, it is much drier, because the winds have lost most of their moisture. The Canterbury Plains, in the "rain-shadow" of the Southern Alps, are famous as a sheep farming area. The farmers also grow wheat, oats and potatoes, as well as fodder crops such as grass and turnips for their sheep.

336

The sheep are sheared at the beginning of summer, which in New Zealand is round about Christmas time. In January the roads across the plains are busy with trucks laden with bales of wool, and with double-tiered lorries taking the sheep to the freezing works.

After the Canterbury sheep have been killed and the carcases frozen, they are exported in ships in refrigerated holds. Lyttelton is the port for Christchurch, main city of Canterbury Province, and is one of several South Island ports which export lamb to Britain.

New Zealand's industries

Most of the industries are connected with agriculture or with making things needed by the people. Many articles needed for the manufacturing industries have to be imported from Britain: cloth, buttons and thread for the clothing industry, and car parts for the motor assembly plants.

The Karapiro Dam is one of many which provides hydro-electricity for the factories of North Island

New Zealand's industries are growing in many places. Electricity, most of it generated from water, provides the power for them.

Taking a trolley load of butter from a churn

SOME INDUSTRIES CONNECTED WITH AGRICULTURE	SOME INDUSTRIES FOR THE NEW ZEALAND MARKET
Meat freezing	Clothing
Butter and cheese making	Car assembly
Ham and bacon curing	Radios
Jam making	Electrical goods
Fruit preserving	Agricultural machinery
Hide tanning	Boots, shoes, saddles
Flour milling	Biscuits
Saw milling	Furniture, printing

New Zealand

The country is fairly prosperous because there is so much agricultural produce to export. The shops stay open until 9 p.m. on Friday evening instead of opening on Saturday. At the weekends, banks, offices, garages and shops are closed. New Zealanders spend much of their leisure out-of-doors; they ski, walk and climb in the mountains of the North and South Islands, fish for trout in the streams, and camp beside the many lakes.

The Southern Alps

A great deal of the South Island is filled by the mountain range of the Southern Alps, with its high peaks which are snow-covered all the year. (The highest is Mt. Cook, 12 349 ft—3763 m.) Sir Edmund Hillary, the New Zealand polar explorer and climber of Everest, did much of his training on the snowfields and glaciers of the Southern Alps.

In winter there are very heavy falls of snow in the Southern Alps. During the summer only the top layer of snow melts, and the water sinks into the snow and freezes underneath the surface. In time a "river of ice", called a glacier, is made. Glaciers move very slowly down the valleys, never more than a

A deep crevasse in Fox Glacier, South Island

few inches a day. Sometimes, as they move, the ice cracks, and huge *crevasses* are formed, many feet wide and deep.

Fiordland

The south-west corner of the South Island, with its mountains, lakes and ocean fiords, has some of the finest scenery in the world. The mountains, which rise steeply from the shores of the lakes and fiords, are covered with thick forests, and crossed by waterfalls and deep river gorges. Parts of the area have not yet been mapped or explored, and a few years ago birds called *notornis*, which the New Zealanders thought were extinct, were found in the forests of fiordland.

A notornis, which was thought to be extinct until naturalists found it again in the forests of fiordland

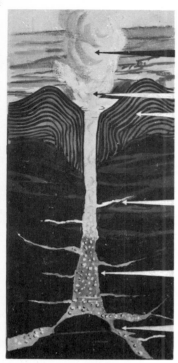

A section through a geyser

Some geysers throw up boiling mud instead of steam and water

Crater of geyser

Deposit of silica

Water flowing through cracks in the rocks makes channels which in time reach the surface and become geysers

Water in lower throat of geyser becomes so hot that it changes to steam. This steam forces the water above into the air

Water meeting hot rocks in the interior of the earth has a temperature much greater than the surface boiling point

Maori guides at Rotorua

Boiling mud at Rotorua

Geysers and hot springs

In the North Island some favourite "sights" are the geysers and hot springs of Rotorua. Here, because there is great heat below the earth's surface, hissing and rumbling can be heard under the ground. In places, there are streams and lakes of hot water, and steaming cliffs and waterfalls.

The Maoris of the hot springs district act as guides to the visitors. There are many "Keep to the Path" notices, for in places boiling fountains of steam and water (called *geysers*) explode high into the air. Everywhere, too, there are pools of creamy mud, which make a continual "plop-plop" noise as the mud bubbles and gurgles.

A cool temperate land: England

A Mediterranean land: Southern France—the grape harvest

Let's remember
The Temperate Lands

(*Temperate* means "moderate" or "mild".)

Remember that *distance from the equator*, the *height of the land*, and *distance from the sea*, all help to decide the climate of a country.

1 The Mediterranean lands

The mild wet winters and warm dry summers of the Mediterranean lands make them ideal countries to live in.

The people of Southern Spain, Italy, Greece and North Africa enjoy the warmth, the clear skies and the vivid colours that are so attractive to northern visitors. In places, however, drought and poor soil cause widespread poverty. Other parts of the world with a Mediterranean climate are California and parts of South Africa and southern Australia. In these lands, many kinds of fruit are grown, especially grapes, olives, oranges, lemons, peaches, apricots,

figs and grapefruit. Wheat thrives, as well as such crops as tobacco and cotton, and rice—where irrigation is possible.

2 The cool temperate lands

Britain is said to have a "cool temperate climate". This means that it is never very hot or very cold. The average temperature is rarely over 70° F (21° C) or below 40° F (4° C) in any one month, and rain falls regularly throughout the year. This is mainly because the winds from the seas in the west cool the land in summer and warm it in winter. Many of the countries of Europe, as well as New Zealand, part of western Canada and the mountainous parts of Chile, are cool temperate lands. In these countries, sheep, cattle, pigs and hens thrive, and so do many grain crops such as wheat, barley and oats. The weather is rarely too extreme to make work difficult, and the cool temperate lands contain some of the most important industrial countries.

Continental grassland: Dakota, U.S.A. Harvest time on the wheat plains

3 The continental grasslands

In the centre of large continents, far from the sea, the summer is usually very hot and the winter very cold. Temperatures often rise above 90° F (32° C) in summer and fall below 0° F (−18° C) in winter. The prairies of central U.S.A. and Canada and the steppes of Russia are the biggest continental grasslands, where crops of wheat, barley, oats and rye are grown during the summer.

South of the equator there are similar lands: in the Argentine pampas, the South African veld, and part of eastern Australia. These are mostly rolling grasslands with few trees. In South Africa and South Australia sheep are reared, and in Argentina cattle are reared and grain is grown. Much wheat is grown in the South Australian grasslands.

4 The deserts of temperate lands

The centres of large areas of land in temperate regions are cut off from rain-bearing winds by distance, and often by mountains as well. These places are deserts, hot in summer and very cold in winter. Such regions are part of central North America, the Gobi desert of Mongolia, Tibet, Iran, and the Patagonian desert of South America. Few people live in these areas, but more may go to them if prospectors find the minerals for which they are searching.

A desert in a temperate land: Tibet. This plateau is 15 000 feet (4600 m) above sea level

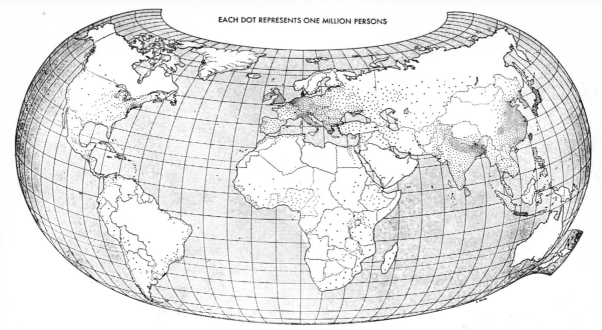

This map shows how the people of the world are spread unevenly over the world's surface. Many people live in the north-east of the U.S.A., in Europe, in India, in China and in Japan

5 East coast lands

The east coasts of the continents of the world have a more extreme climate than the west coasts, with hot summers and cool

An East Coast land: Vermont, U.S.A. Vermont is often called the "four-season State", because of the great contrasts between its seasons

or cold winters. (Most winds are west winds. In the west of the continents they blow from the sea; in the east they blow from the land, which is hotter in summer and cooler in winter.)

Eastern Canada and the east of the U.S.A., Japan, Natal and the south-east of Australia are east coast lands.

Nearer the equator, as in South China and Queensland, heavy summer rains allow rice and sugar cane to be grown. Cotton is grown in the south-east of the U.S.A., and coffee in Brazil.

In Manchuria and eastern Canada the winters are much colder, and rivers, such as the St. Lawrence, are frozen in winter. There is less rain, but mixed farming and dairy farming are carried on.

In many east coast lands, farmland is giving way to factories.

342

THE COLD LANDS AND SEAS
18 The Northern Lands and Seas

The tundra

Huge plains, the *tundra*, stretch for thousands of miles across the north of Russia, Scandinavia, Canada and Alaska. Here, the long, dark winters are bitterly cold, and the rivers flow for only about three months in the year. The sea is ice-covered except for a few weeks in the summer.

Very few people live in these lands. In the past, they earned a living mainly by fishing, hunting and keeping herds of reindeer. Today they are learning to help in the search for minerals, and are trading furs for rifles, radios and tins of food.

How glad the people of the Arctic are when their three months' summer comes. For many days the sun never sets, the temperature may rise to 100° F (38° C), and in places the moss-covered plains are carpeted with small flowers. The air is alive with the sound of birds, and brightly coloured moths fly over the Arctic meadows.

But even in summer it is difficult to travel across much of the *tundra*. A few feet below the surface the earth is still frozen, and when the water from the melting rivers overflows, it cannot soak into the ground. The marshes formed in summer by these melting rivers are the breeding grounds of swarms of mosquitoes, which make life unbearable for any traveller who tries to go through the swamps.

Left: *A reindeer*
Middle: *A plane with skis*
Right: *A mosquito*

343

The Alaska Highway

The Alaska Highway stretches for 1523 miles, from Dawson Creek in British Columbia to Fairbanks in Alaska

Alaska

A hundred years ago the territory of Alaska, in North America, was so unimportant that it was sold by Russia to the United States for less than two million pounds. Today it is one of the most valuable states of the U.S.A. In area it is ten times as large as England and Wales, yet only 234 000 people live there.

Fishing and mining make the wealth of Alaska. The rivers teem with salmon, which are tinned and sent to other countries. Off the coast, cod, herring and sardines are caught. Quick-freeze plants have been built so that fresh fish can be sent to the mainland of the U.S.A.

Gold, copper and platinum are the most important minerals of Alaska, but coal and other minerals are also mined. Forestry and fur trapping are very important industries.

Alaska has a very varied climate. Warm south-west sea currents give southern Alaska a mild climate. Most of the people live there, rather than in the cold interior or in the north. All the towns are in the south: Anchorage, Fairbanks, Ketchikan, and Juneau the capital. They are modern towns with blocks of flats, electricity, super-markets and hotels. But life is expensive, for nearly all the necessities of life, and all the luxuries, have to be brought by ship or plane from the mainland of the U.S.A.

Because Alaska has become so important in recent years, a road has been built from the U.S.A., through Canada, to Alaska. The road is called the Alaska Highway. It is 1523 miles (2450 km) long, and is open all the year round, though motorists take spare parts, food and extra clothes in case they break down when they are miles from civilisation.

In the far north helicopters are used for survey work

A general view of a gold mine at Yellowknife

Canada

Much of northern Canada is a country of lakes and coniferous trees, but in the far north it is too cold even for trees. The native people are Eskimos and North American Indians who, in the past, made a living by hunting and trapping animals for their furs, including the otter, musk rat, white fox, beaver, marten and ermine.

Nowadays the Eskimos and Indians are learning how to become miners and mechanics, to work with drilling machines, grabs and caterpillar tractors. Gold has been found near Yellowknife, and silver and pitchblende (from which uranium is obtained) near the Great Bear Lake. Oil, lead and zinc have also been found.

There are few roads and railways in this part of Canada and planes and helicopters are used for much of the survey work which is necessary in the search for minerals. The planes are often fitted with skis, so that they can land on the frozen snow.

The more prosperous Eskimos of the Canadian Arctic own Peterhead cutters, fitted with petrol engines and auxiliary sail

This Russian ice-breaker is atomic powered. It can stay at sea for two to three years, since it uses very little fuel. As the propellers force the ship forward, the sloping bows are pushed up over the ice, and the weight of the ship breaks the ice, so making a way through

"Opening up" the north of Canada

Roads are being built, and the 385-mile Mackenzie Highway now runs from Grimshaw to Hay River. Roadmaking is difficult because the ground is frozen solid for most of the year, and roads tend to crack when the ground thaws. Rivers are being dammed to provide hydro-electric power.

A petroleum exploration team in the northern forests of Canada

Northern Russia and Siberia

Today, in some parts of the far northern wastes of Russia, minerals such as coal, oil and gold are mined. New towns have been built, with power stations to generate electricity. The Government wants people to make their homes in these parts, but first there must be food for them to eat. Russian scientists are experimenting with different varieties of grain, fruit and vegetables. Crops of sugar-beet, wheat and oats are being grown.

The Northern seas

There are no regular shipping routes to the north of Russia and Canada. The north-west passage (the route round the north of Canada), and the north-east passage (north of Russia), are frozen up for most of the year. Each country has ice-breaker ships which force a way through the summer ice, helped by helicopters which look for weaknesses in the ice.

346

The seas off Alaska and Norway are warmed by south-west sea currents, and so do not freeze in the summer.

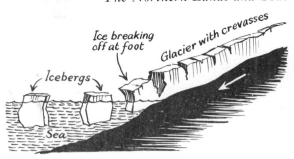
How icebergs are formed

The North Pole

In the far north of the world, around the place we call the North Pole, there is no land, but only ice. In places, the ice is as much as two miles thick, yet it is never so thick that it rests on the sea bed; it is always floating. (This was proved when the American atomic submarine *Nautilus* sailed right under the Arctic ice in 1958.)

Scientists have set up research stations on the polar ice, with huts, wireless sending and receiving sets, tractors and all kinds of modern equipment. They are studying the thickness and drift of the ice, the depth of the sea beneath it, and the weather in Polar regions.

Icebergs

In the northern seas are many icebergs which have broken off the edge of the polar ice, or off the ends of glaciers. They are very dangerous to shipping, and from March to June each year, ships and aircraft of the International Iceberg Patrol keep watch for icebergs in the North Atlantic Ocean.

Arctic air routes

Airfields have been built as far north as Thule, in northern Greenland, where there is a large American base. Today there are regular passenger services by jet air-liner from Paris to Tokyo, crossing the polar ice. The journey takes 17 hours. Piston-engined planes on the usual route take 27 hours. Other Arctic routes are planned, since they are much shorter than the old routes.

The "Great Circle" route from London to San Francisco crosses Greenland. Similarly, the quickest way from Europe to Japan is over the North Pole

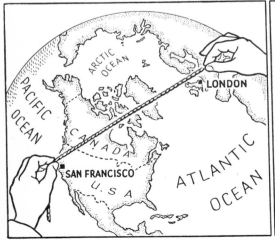

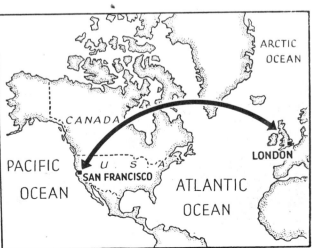

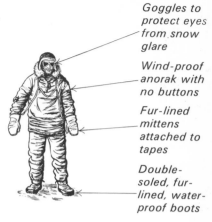

A Sno-cat on the edge of a deep crevasse

Goggles to protect eyes from snow glare

Wind-proof anorak with no buttons

Fur-lined mittens attached to tapes

Double-soled, fur-lined, water-proof boots

An explorer's dress in the Antarctic

19 The Antarctic

Why do men make expeditions to the loneliest and most dangerous places on earth? In the past most explorers were seeking gold, or new opportunities for trade, but modern explorers usually want the satisfaction of reaching distant places where nobody has been before.

Some explorers are scientists trying to find out more about the earth we live in. When Vivian Fuchs made the first crossing of Antarctica, the land of the South Pole, many of the men in his party were geologists or weather experts.

The Antarctic

Antarctica is a bleak barren land, almost as large as the U.S.A. and Australia together. It is much colder and higher than the Arctic, and no plants or animals live there, except for a few penguins and seals near the sea.

The centre of Antarctica is a flat plateau 10 000 ft (3000 m) above sea level, and the whole land is perpetually covered by ice up to 8000 ft (2500 m) thick.

Although the Antarctic is difficult to explore, several nations have sent expeditions to claim sections of the land. Later, other parties were sent to carry out research and to set up weather stations.